An Operator's Guide to Biological Nutrient Removal (BNR) in the Activated Sludge Process

Michael H. Gerardi

Illustrations by Brittany Lytle

An Operator's Guide to Biological Nutrient Removal (BNR) in the Activated Sludge Process

ISBN: 978-0-8206-0416-9
eBook ISBN: 978-0-8206-0417-6

© Chemical Publishing Company, Inc. - 2016

Chemical Publishing Company:
www.chemical-publishing.com

Printed in the United States of America

to Jack Ryan Stabley Jr.

Preface

Biological nutrient removal (BNR), the removal of nitrogen and phosphorus from wastewater, is a complex process. Although the activated sludge process is an efficient technology for the removal of biochemical oxygen demand (BOD) and total suspended solids (TSS), it provides less-than-optimal conditions for the removal of nitrogen and phosphorus, and presents numerous challenges to the operator trying to satisfy the many requirements for several different groups of bacteria. In addition to satisfying the requirements there are numerous, highly variable operational conditions that impact BNR. These conditions include: changes in strength and composition of the wastewater, alkalinity and pH, temperature, and presence of inhibitory and toxic wastes. Even fluctuations in flows, especially from inflow and infiltration, can adversely impact the aerobic, anoxic, and anaerobic conditions needed for successful BNR.

Of the three treatment processes, nitrification, denitrification, and enhanced biological removal, nitrification is often the most difficult to achieve. Therefore, a large portion of this book reviews nitrification. Operators of the activated sludge process need to understand the basic biological, chemical, and physical requirements for BNR in order to improve the performance of these treatment processes. *An Operator's Guide to Biological Nutrient Removal (BNR) in the Activated Sludge Process* is intended to help operators in the monitoring, troubleshooting, and process control of BNR. Numerous tables and figures are included in the book to help the operator understand the biological and chemical reactions that are involved in BNR processes and how the reactions can be monitored for process control.

   Design of BNR processes is not addressed in this book. Design is addressed in numerous engineering publications. The book serves to help operators achieve permit compliance for nitrogen and phosphorus discharge limits and obtain cost-effective operation.

Michael H. Gerardi
Linden, Pennsylvania

Contents

List of Tables   ix
List of Figures   xiii

PART ONE: NITRIFICATION   1

Chapter 1 Introduction   2

Chapter 2   Nitrogenous and Phosphorous Compounds   37

Chapter 3 Nitrification: The Basics   51

Chapter 4 Nitrifying Bacteria   58

Chapter 5 Nitrification and Limiting Factors   81

Chapter 6 Promoting Nitrification   89

PART TWO: DENITRIFICATION   90

Chapter 7 Denitrification: The Basics   102

Chapter 8 Denitrifying Bacteria   106

Chapter 9 Denitrification and Limiting Factors   125

PART THREE: BIOLOGICAL
                 PHOSPHORUS REMOVAL   126

Chapter 10 Biological Phosphorus Removal:
                 The Basics   139

Chapter 11 EBPR: Process Control   149

Abbreviations and Acronyms   151
Glossary   154
Bibliography   158

List of Tables

Chapter 1

1.1	Industrial Wastewaters that Contain Nitrate (NO_3^-) or Nitrite (NO_2^-)	8

Chapter 2

2.1	Major Nitrogenous Compounds	14
2.2	Oxidation States of Nitrogen	15
2.3	Molecular Weight of Ammonia (NH_3)	17
2.4	Molecular Weight of Ammonium (NH_4^+)	20
2.5	Industrial Discharges Containing Significant Quantities of Ammonium	22
2.6	Molecular Weight of Nitrite (NO_2^-)	27
2.7	Molecular Weight of Nitrate (NO_3^-)	28
2.8	Industrial Wastewaters that Contain Nitrate (NO_3^-) or Nitrite (NO_2^-)	29
2.9	Quantity of Nitrogen in 1 mg/L in Reported Test Values	31
2.10	Major Phosphorous Compounds	32
2.11	Molecular Weight of Orthophosphate (HPO_4^{2-})	33
2.12	Coagulants Used to Precipitate Orthophosphate	33
2.13	Quantity of Phosphorus in 1 mg/L in Reported Test Values	36

Chapter 3

3.1	Sources of Ammonium, Nitrite, and Nitrate	38

Chapter 4

4.1	**Characteristics of *Nitrosomonas* and *Nitrobacter***	52
4.2	**Differences between Autotrophic Nitrifying Bacteria and cBOD-degrading Heterotrophic Bacteria**	53
4.3	**Favorable Operational Factors for Nitrification in the Mixed Liquor**	56
4.4	**Guideline MCRT Needed for Nitrification in Temperate Regions of North America**	57

Chapter 5

5.1	**Alkali Compounds Commonly used for Alkalinity Addition**	61
5.2	**Guideline ORP Values and Cellular Activity**	67
5.3	**Examples of Inhibitory or Toxic Compounds or Ions to Nitrifying Bacteria**	69
5.4	**Alcohols that are Recognizable, Soluble cBOD Compounds**	72
5.5	**cBOD:TKN Ratio and Fraction of Nitrifying Bacteria in MLVSS**	74
5.6	**Side Stream Processes (Recycle Streams) that have Nitrogen and Phosphorus**	74
5.7	**Checklist for Identifying Temperature and Limiting Factors Responsible for the Occurrence of Incomplete Nitrification**	78

Chapter 6

6.1	**Values or Ranges of Values for Parameters for Promoting Nitrification**	82
6.2	**Operational Measures to Promote Nitrification**	82

Chapter 7

7.1	**Filamentous Organisms that can be controlled in an Anoxic Condition**	92
7.2	**Organic Compounds Commonly Used as a Carbon Source for Denitrification**	96

Chapter 8

8.1	Significant Differences in Habitat, Metabolism, and Roles Performed in the Activated Sludge Process	103
8.2	Comparison of Aerobic Bacteria and Facultative Anaerobic Bacteria to Degrade cBOD	104

Chapter 9

9.1	Effects of Nitrification and Denitrification upon the Activated Sludge Process	106
9.2	Respiration and Sludge Yield	112
9.3	Guideline ORP Values and Cellular Activity	115
9.4	Comparison of Energy Obtained and Sludge Yield When Denitrifying Bacteria use Oxygen or Nitrate to Degrade cBOD	116
9.5	Examples of Highly Soluble, Easily Assimilated Carbon Sources for Denitrification	119

Chapter 10

10.1	Operational Measures Available for Phosphorus Removal	126
10.2	Aluminum and Iron Salts used to Precipitate Phosphate	127
10.3	Nutrient Removal Processes	133
10.4	Fermentative or Acid-forming Bacteria	134
10.5	Poly-P Bacteria	135
10.6	Volatile Fatty Acids	135

Chapter 11

11.1	Treatment Processes Available to Satisfy TP Discharge Requirements	139
11.2	Commonly Recommended Values for EBPR	141
11.3	Operational Conditions Associated with Poorly Settling Solids	145
11.4	Operational Conditions Associated with the Production of Fine Solids	145
11.5	Growth Ranges for Thermophilic, Mesophilic, and Psychrophilic Bacteria	146

List of Figures

Chapter 1

1.1	Activated sludge process	2
1.2	Reactors used for nitrification and denitrification	5
1.3	Sequential nitrification/denitrification	6
1.4	Oxygen gradient	9
1.5	Reactors used for biological phosphorus removal	11
1.6	Sequential biological phosphorus release/biological phosphorus uptake	12

Chapter 2

2.1	Atomic symbols, numbers, and masses	16
2.2	Commonly observed protozoa, rotifers, and soil nematodes in the activated sludge process	19
2.3	Methylamine, dimethylamine, and trimethylamine	21
2.4	Substituted ammonium compounds	22
2.5	Hydrolysis of urea	24
2.6	Basic structure of amino acids and examples of amino acids	25
2.7	Examples of amines having substituted hydrogen atoms	25
2.8	Deamination	30
2.9	Groups of nitrogenous compounds	34
2.10	Groups of phosphorous compounds	35
2.11	Hydrolysis of polyphosphate	35
2.12	Polyphosphate	36
2.13	Phosphate	36

Chapter 3

3.1   Inhibition of nitrification in the sewer system   39
3.2   Aerobic digester and nitrogenous compounds   43
3.3   Anaerobic digester and nitrogenous compounds   44
3.4   One-stage and two-stage nitrification processes   46
3.5   Indicators of nitrification across an aeration tank   47
3.6   Duckweed   48
3.7   *Epistylis*   49

Chapter 4

4.1   In-folding of cell membrane in nitrifying bacteria   51
4.2   Dependency of NOB upon AOB for its energy substrate   54
4.3   Relationship between fraction of nitrifying bacteria and changes in cBOD-to-TKN ratio   55

Chapter 5

5.1   Temperature and its impact upon nitrification   59
5.2   Calculating alkalinity needed for nitrification   64
5.3   Plug-flow mode of operation   71
5.4   Concentration of ammonium in mixed liquor influent and mixed liquor   72

Chapter 6

6.1   Dissolved oxygen profile and stratification   83
6.2   Rotating aeration tanks in-line and off-line   87

Chapter 7

7.1   Removal of nitrogen from wastewater   92
7.2   Removal of electrons by nitrate   94
7.3   Oxygen gradient   95
7.4   Denitrification in an aeration tank   97
7.5   Indicators of denitrification across a secondary clarifier   98
7.6   Rise times in a settleometer   99
7.7   Calculating alkalinity produced through denitrification   101

Chapter 9

9.1	Sequential nitrification/denitrification	107
9.2	Assimilatory and dissimilatory nitrate reduction	108
9.3	Bacterial respiration and fermentation in the sewer system	110
9.4	Respiration and fermentation	111
9.5a	Problematic conditions caused by bacterial respiration using nitrate in the sewer system	113
9.5b	Problematic conditions in the treatment plant caused by influent nitrite and nitrate	114
9.6	Simultaneous use of oxygen and nitrate by heterotrophic bacteria	117
9.7	Absorption and adsorption of cBOD	118
9.8	Endogenous decay	120
9.9	Temperature and its impact upon denitrification	122
9.10	Modified Ludzack-Ettinger process	123
9.11	Four-stage Bardenpho process	124

Chapter 10

10.1	Mainstream and side stream processes for the removal of phosphorus	130
10.2a	Nutrient removal processes for the removal of phosphorus or nitrogen	131
10.2b	Nutrient removal processes for the removal of phosphorus or nitrogen	132
10.3	Biological phosphorus release	137
10.4	Biological phosphorus uptake	138

PART ONE: NITRIFICATION

Chapter 1

Introduction

The activated sludge process is a biological wastewater treatment process that uses bacteria to treat municipal wastewater and numerous industrial wastewaters. The activated sludge process typically consists of an aeration tank and a downstream clarifier (Figure 1.1). It is the most commonly used technology for the treatment of municipal wastewater and industrial wastewaters. The bacterial solids or floc particles in the aeration tank are maintained in suspension through mechanical action or aeration, while the floc particles settle in the clarifier which has a nearly quiescent condition.

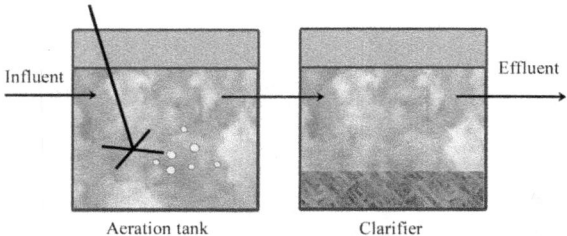

Influent   Effluent

Aeration tank   Clarifier

Figure 1.1 Activated sludge process: The activated sludge process consists of at least two tanks or groups of tanks. The first tank or upstream tank is the aeration tank containing the mixed liquor. Wastewater enters the aeration tank where it is treated by a large consortium of bacteria contained in suspended floc particles. After treatment the mixed liquor leaves the aeration tank and enters the second tank or downstream tank. The downstream tank is the clarifier. Because the activated sludge process provides secondary treatment, the clarifier is referred to as the secondary clarifier. Here, a rather quiescent condition occurs that permits the settling of the floc particle or solids. Once settled, the solids may be returned to the aeration tank or wasted from the system. Primary treatment upstream of the activated sludge process is physical treatment (screening, settling, etc.) of wastewater.

Wastes that are treated include: 1) organic compounds that exert an oxygen demand (carbonaceous, biochemical oxygen demand or cBOD), 2) nitrogenous compounds that exert an oxygen demand (nitrogenous, biochemical oxygen demand or nBOD), 3) suspended solids (total suspended solids or TSS), 4) inorganic wastes such as heavy metals, and 5) pathogens. By degrading or removing these wastes the health of the community is provided a large measure of protection, and the adverse impact of wastewater upon the receiving body of water is greatly reduced.

Today, more and more activated sludge processes are required to remove or reduce the quantity of nitrogen (N) and the quantity of phosphorus (P) that are discharged to a receiving body of water. Therefore, municipal wastewater treatment plants are required to nitrify to satisfy an ammonia discharge limit or nitrify and denitrify in order to satisfy a total nitrogen discharge requirement. Plants nitrify and denitrify using autotrophic nitrifying bacteria and heterotrophic cBOD-removing bacteria, respectively. Plants that practice enhanced biological phosphorus removal (EBPR) to satisfy a total phosphorus discharge requirement also use two different groups of bacteria: heterotrophic acid-forming (fermentative) bacteria and heterotrophic phosphorus-accumulating bacteria.

The differences between growth requirements for autotrophic (nitrifying) bacteria and heterotrophic (cBOD-degrading) bacteria underscore the complexity of managing different bacterial populations in order to satisfy discharge requirements for cBOD, ammonia, total nitrogen (TN), and total phosphorus (TP). Resolving these differences is a significant challenge to the successful operation of the activated sludge process because numerous parameters affect the activity and changes in dominant groups of bacteria.

The requirements for nitrogen and phosphorus or nutrient removal are to prevent undesired impact of nutrients upon receiving waters and to protect the health of the community. Nitrogen and phosphorus cause many undesired environmental impacts in receiving waters including: 1) blooms of aquatic plants, especially algae, 2) dissolved oxygen depletion, 3) eutrophication, and 4) toxicity to aquatic organisms. For examples, highly nitrified effluents can accelerate algal blooms, and nitrogen in the form of nitrate (NO_3^-) in potable water is responsible for methemoglobinemia or "blue baby" disease.

Nitrogen

Nitrogen is removed from wastewater by several measures. It may be stripped to the atmosphere at high pH as ammonia (NH_3). It is assimilated into new bacterial growth (sludge production), and it may be oxidized (nitrified) and then reduced (denitrified). Of these removal measures, biological nitrification coupled with biological denitrification is the major measure for removing nitrogen from the wastewater (Figure 1.2). Biological removal of nitrogen can be performed by several treatment configurations using single-stage processes having various treatment zones or separate-stage processes having separate treatment units for nitrification and denitrification. Whatever configuration is used, it requires an aerobic condition for converting ammonium to nitrate (nitrification) and an anoxic condition for converting nitrate to insoluble molecular nitrogen. Complete nitrogen removal (nitrification and denitrification) is possible if the chemical oxygen demand (COD)-to-total nitrogen ratio is ≥ 10.

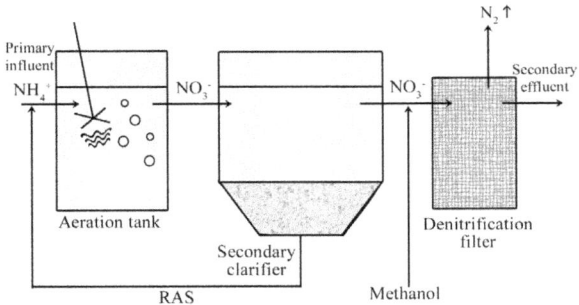

Figure 1.2 Reactors used for nitrification and denitrification: There are two reactors that are used for nitrification and denitrification. The first reactor is an aeration tank where cBOD is oxidized and ammonium is oxidized to nitrate. Nitrate leaves the aeration tank and passes through the secondary clarifier in the waste stream to a denitrification tank or filter. The filter has either no dissolved oxygen or a relatively small quantity of dissolved oxygen. However, little soluble cBOD is in the wastewater entering the filter. The cBOD has already been oxidized in the aeration tank. Therefore, in order to have the bacteria in the filter use nitrate for respiration, cBOD or a carbon source like methanol is added to the filter. Once added to the filter, bacteria degrade the carbon using any residual dissolved oxygen. After the residual dissolved oxygen has been depleted, carbon is degraded by nitrate entering the filter.

Nitrification is the autotrophic oxidation of ammonium or ionized ammonia (NH_4^+) to nitrite (NO_2^-) and then nitrate (NO_3^-). Nitrification does not remove nitrogen from a waste stream. Denitrification reduces nitrate to gaseous molecular nitrogen or dinitrogen gas (N_2). As insoluble molecular nitrogen, nitrogen leaves the waste stream and enters the atmosphere. When nitrification and then denitrification are coupled together or operated as "sequential nitrification/denitrification," nitrogen is removed from the waste stream (Figure 1.3).

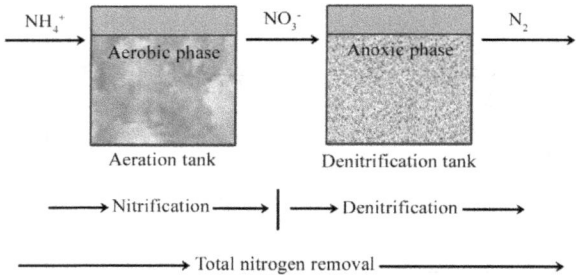

Figure 1.3 Sequential nitrification/denitrification: Sequential nitrification/ denitrification consists of two reactors in plug-flow mode of operation. The first reactor may be the aeration tank, and the second reactor may be the denitrification tank. Denitrification occurs in a reactor having an anoxic condition, absence of dissolved oxygen, and presence of nitrate and soluble cBOD. In plug-flow mode of operation the first reactor may be a denitrification tank having mixing action but no aeration and the second reactor the aeration tank. Some of the nitrate produced in the aeration tank can be recycled back to the denitrification tank. In the denitrification tank, bacteria use recycled nitrate to degrade influent soluble cBOD.

Although the processes of nitrification and denitrification seem fairly straightforward, successful nitrification and denitrification are often difficult to achieve due to changes in operational conditions. The most challenging task for treatment plant operators is balancing operational conditions to promote the acceptable activity of abundant populations of nitrifying bacteria and heterotrophic (denitrifying) bacteria.

Nitrification

Nitrification occurs in an activated sludge process for three reasons: 1) the process is required by a regulatory agency to satisfy an ammonia or total nitrogen discharge requirement, 2) the process is not required to satisfy a discharge requirement but is operated to produce nitrate, and 3) the process is not required to satisfy a discharge requirement, and nitrification is not desired; but the operational conditions are suitable for nitrification, and the process "slips" into nitrification. Processes that are operated to nitrify even though nitrification is not required do so in order to produce nitrate that 1) indicates a stable and acceptable operational condition and 2) is used to produce an anoxic condition for the control of undesired filamentous organism growth.

Nitrification occurs naturally in the soil as well as an activated sludge process. Nitrification is the biochemical (bacterial) oxidation of ammonium (NH_4^+) to nitrite (NO_2^-) (Equations 1.1), or the biochemical (bacterial) oxidation of nitrite to nitrate (NO_3^-) (Equations 1.2). Although each biochemical reaction can occur independently of the other typically, the biochemical reactions usually are coupled, that is, they occur together (Equations 1.3). In municipal activated sludge processes, ammonium is present in domestic wastewater and therefore the influent, but nitrate and nitrite are not, unless there is an industrial discharge of nitrate or nitrite to the sewer system (Table 1.1).

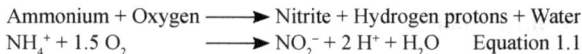

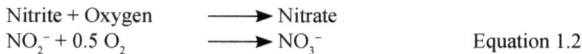

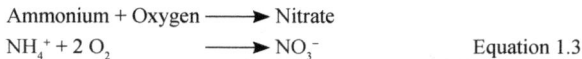

Ammonium + Oxygen $\longrightarrow$ Nitrite + Hydrogen protons + Water
$$NH_4^+ + 1.5\,O_2 \longrightarrow NO_2^- + 2\,H^+ + H_2O \qquad \text{Equation 1.1}$$

Nitrite + Oxygen $\longrightarrow$ Nitrate
$$NO_2^- + 0.5\,O_2 \longrightarrow NO_3^- \qquad \text{Equation 1.2}$$

Ammonium + Oxygen $\longrightarrow$ Nitrate
$$NH_4^+ + 2\,O_2 \longrightarrow NO_3^- \qquad \text{Equation 1.3}$$

Table 1.1 Industrial Wastewaters That Contain Nitrate (NO_3^-) or Nitrite (NO_2^-)

Industrial Discharge	Presence of Nitrite (NO_2^-)	Presence of Nitrate (NO_3^-)
Corrosion inhibitors	X	
Leachate (biologically pretreated)	X	X
Meat processing (preservatives)	X	
Meat processing (biologically pretreated)	X	X
Meat processing (flavor additives)		X
Steel mill	X	X

Denitrification

Denitrification refers to the biological process of removing nitrogen from wastewater. It is the conversion of nitrate and nitrite to insoluble nitrogenous gases, molecular nitrogen (N_2), and nitrous oxide (N_2O). Denitrification occurs in an activated sludge process for three reasons: 1) the process is required by a regulatory agency to satisfy a total nitrogen discharge limit, 2) the process requires nitrates in order to produce an anoxic condition in an anoxic selector to control the undesired filamentous organism growth, and 3) operational conditions in a reactor such as a secondary clarifier are suitable for denitrification. Undesired denitrification is also known as "clumping" or "dark sludge rising."

Denitrification occurs in the absence of dissolved oxygen or the presence of an oxygen gradient (Figure 1.4). During denitrification, oxygen in nitrate and nitrite is gradually removed from the nitrogen by denitrifying bacteria through a series of biochemical reactions. The series include the reductions of nitrate to nitrite to nitric oxide (nitrogen monoxide) to nitrous oxide (nitrogen dioxide or laughing gas) to molecular nitrogen (dinitrogen) (Equation 1.4). Nitrogen is removed from wastewater when insoluble nitrous oxide and molecular nitrogen are released to the atmosphere. Nitrogen removal is often and easily combined with enhanced biological phosphorus removal in the activated sludge process.

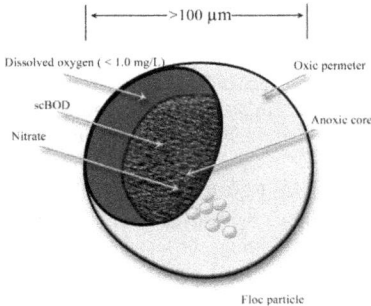

Figure 1.4 Oxygen gradient: Denitrification can occur in the presence of measureable dissolved oxygen. This can occur if the bulk solution contains nitrate, soluble cBOD, and < 1.0 mg/L dissolved oxygen, and the floc particle is > 100 μm in diameter. Bacteria on the perimeter of the floc particle use dissolved oxygen to degrade the cBOD. Because the relatively small quantity of dissolved oxygen is depleted as it moves toward the core of the floc particle and nitrate is not, nitrate is used in the core of the floc particle to degrade cBOD. In activated sludge processes that nitrify, a relatively high MCRT is needed to establish an abundant population of slow-growing nitrifying bacteria. The high MCRT promotes the growth of filamentous organisms. The organisms in turn provide strength to the floc particle, allowing the particle to overcome turbulence and grow in size.

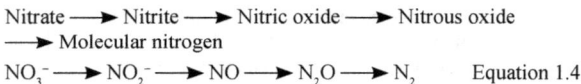

Nitrate ⟶ Nitrite ⟶ Nitric oxide ⟶ Nitrous oxide
⟶ Molecular nitrogen

$$NO_3^- \longrightarrow NO_2^- \longrightarrow NO \longrightarrow N_2O \longrightarrow N_2 \qquad \text{Equation 1.4}$$

Phosphorus

Phosphorus is also removed from the wastewater by several measures. It is assimilated into new bacterial growth (sludge production). It may be absorbed and stored in bacterial cells as insoluble phosphorus or volutin granules, and it may be precipitated from solution. Of the biological measures for removing phosphorus, its absorption and storage as insoluble phosphorus granules (EBPR) or luxury uptake of phosphorus is a major operational measure for removing phosphorus from the wastewater. Biological phosphorus removal like the removal of nitrogen requires two events, biological phosphorus release and then biological phosphorus uptake. In some activated sludge processes phosphorus is removed from wastewater by a combination of EBPR and chemical precipitation.

Enhanced biological phosphorus removal

Enhanced biological phosphorus removal (EBPR) requires two groups of bacteria, acid-forming or fermentative bacteria and phosphorus-accumulating bacteria. Phosphorus-accumulating bacteria are also known as poly-P bacteria and are one of several phosphorus-accumulating organisms (PAO) (Table 1.2). EBPR also requires the use of two reactors (Figure 1.5). The first reactor has an anaerobic/fermentative condition where fatty acids are produced by fermentative bacteria. Here, biological phosphorus release occurs. The second reactor has an aerobic condition where biological phosphorus uptake occurs. When biological phosphorus release and biological phosphorus uptake are coupled together or operated as "sequential biological phosphorus release/biological phosphorus uptake," phosphorus is removed from the waste stream by wasting secondary solids (bacteria) that are rich in insoluble phosphorus (Figure 1.6).

Table 1.2 Phosphorus-Accumulating Organisms

Algae
Bacteria including Actinomycetes
Cyanobacteria
Fungi including yeast and filamentous fungi
Protozoa

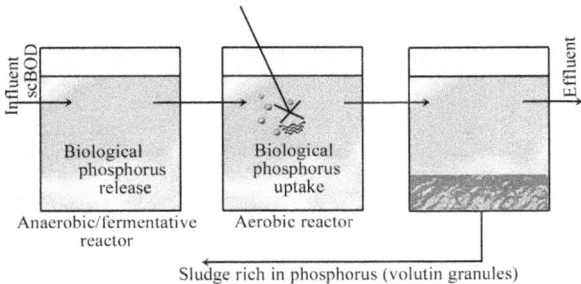

Figure 1.5 Reactors used for biological phosphorus removal: Biological phosphorus removal is achieved using two reactors in plug-flow mode of operation. The first reactor or upstream reactor is an anaerobic/fermentative reactor for biological phosphorus release. The second reactor is an aerobic reactor for biological phosphorus uptake.

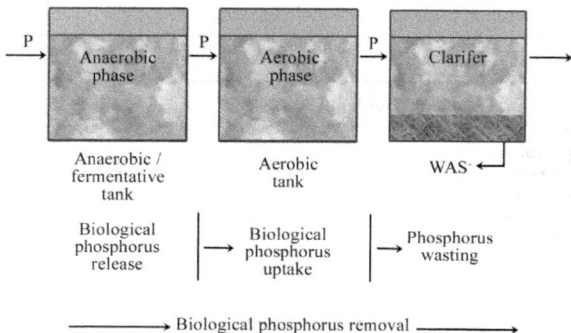

Figure 1.6 Sequential biological phosphorus release/biological phosphorus uptake: Enhanced biological phosphorus removal requires the use of two reactors in plug-flow mode of operation. The first reactor or upstream reactor is operated as an anaerobic/fermentative reactor where highly soluble, volatile fatty acids are produced and cellular phosphorus is released by phosphorus-accumulating bacteria or Poly-P bacteria. The second reactor or downstream reactor is operated as an aerobic reactor where Poly-P bacteria absorb or uptake phosphorus.

Chapter 2

Nitrogenous and Phosphorous Compounds

There are several nitrogenous (nitrogen-containing) and phosphorous (phosphorus-containing) compounds that must be monitored and regulated in order to ensure discharge permit compliance for nutrient removal. Nitrogenous compounds discharged from the activated sludge process can have many harmful environmental effects including: 1) adverse public health effects including methemoglobinemia, 2) dissolved oxygen depletion in the receiving body of water, 3) reduction in chlorine disinfection efficiency, 4) reduction in options for receiving water reuse, and 5) toxicity to aquatic life. Phosphorous compounds discharged from the activated sludge process contribute to the excess growth of aquatic plants, especially algae.

The nitrogenous compounds of concern include: 1) ammonia, 2) ammonium, 3) amine, 4) organic nitrogen, 5) nitrite, 6) nitrite oxide 7) nitrate, 8) nitrous oxide, and 9) molecular nitrogen (Table 2.1). These nine compounds have six possible oxidation states for nitrogen (Table 2.2). However, only nitrogen in the −3 oxidation state is used by bacteria for the nitrogen nutrient for cellular synthesis of nitrogenous compounds such as amino acids, proteins, enzymes, and genetic material. Organic nitrogen compounds must be degraded and the amine group must be removed from the organic nitrogen compound. Because ammonia is toxic, nontoxic ammonium is the primary nitrogen nutrient. In the absence of ammonium, bacterial cells will use nitrate as the nitrogen nutrient. Although

nitrogen in nitrate is in the +5 oxidation state, bacterial cells absorb
nitrate, remove the oxygen atoms, and add hydrogen atoms to the
nitrogen thus converting nitrogen to the −3 oxidation state from the
+5 oxidation state.

Table 2.1 Major Nitrogenous Compounds

Compound	Formula	Role(s) Performed
Ammonia	NH_3	Toxic with increasing pH
Ammonium	NH_4^+	Primary nitrogen nutrient Energy substrate for ammonium-oxidizing bacteria
Amine	$-NH_2$	Forms ammonia/ammonium depending on pH
Organic nitrogen	CH_2ON*	Carbon and hydrogen containing compounds having an amine group, examples include amino acids and proteins
Nitrite	NO_2^-	Intermediate compound produced during nitrification Intermediate compound produced in denitrification Energy substrate for nitrite-oxidizing bacteria Acts as the "chlorine sponge" Toxic at low concentrations
Nitrate	NO_3^-	Final (most oxidized) compound produced in nitrification Secondary nitrogen nutrient Toxic at high concentrations
Nitric oxide	NO	Intermediate compound produce during nitrification Toxic at low concentrations

Compound	Formula	Role(s) Performed
Nitrous oxide	N_2O	Intermediate compound produced in denitrification
		Insoluble gas released to the atmosphere in denitrification
Molecular nitrogen	N_2	Final compound produced in denitrification

*CH_2O designates an organic nitrogen compound having carbon, hydrogen, oxygen, and nitrogen.

Table 2.2 Oxidation States of Nitrogen

Compound	Formula	Oxidation State of Nitrogen
Ammonia	NH_3	−3
Ammonium	NH_4^+	−3
Amine	−NH	−3
Organic nitrogen	CH_2ON	−3
Nitrite	NO_2^-	+3
Nitrate	NO_3^-	+5
Nitric oxide	NO	+2
Nitrous oxide	N_2O	+1
Molecular nitrogen	N_2	0

Nitrogen

Nitrogen exists in many different forms in wastewater. Major sources of nitrogen include untreated sewage, industrial wastewater, food processing, and septic tank waste. Nitrogen enters wastewater mostly from humans in urine as urea (NH_2COH_2N). Untreated sewage in a municipal wastewater treatment plant has total nitrogen concentrations ranging from 20 to 85 mg/L. The majority of the nitrogen is found in ammonium. Nitrogen in domestic wastewater consists of approximately 60% ammonium-nitrogen and approximately 40% organic nitrogen. About one-fifth of the BOD in domestic wastewater is due to total nitrogen (TN).

Activated sludge processes decrease the quantity of total nitrogen entering the process through bacterial growth (sludge production) and solids removal. However, bacterial growth and solids removal alone allow much of the ammonium-nitrogen to leave the process in the effluent.

Nitrogenous compounds

Ammonia

Ammonia (NH_3) is a compound of nitrogen and hydrogen. It is a colorless gas and has an irritating, pungent odor. The molecular weight (molar mass or mol) of ammonia is 17 g/mol (Figure 2.1) (Table 2.3). Nitrogen makes up 82% of the weight of ammonia. Therefore, an analytical test value of 1 mg/L of NH_3 is the same as 0.82 mg/L of NH_3-N.

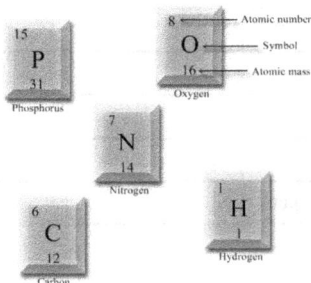

Figure 2.1 Atomic symbols, numbers, and masses: The atomic chart of the elements or periodic table is a tabular arrangement of the chemical elements

ordered by their atomic number. The chart is used to determine the weight of an element in any compound. Each element on the chart has an atomic symbol to identify the element: P for phosphorus, O for oxygen, N for nitrogen, C for carbon, and H for hydrogen. Each element also has the element's atomic number in the upper lefthand corner and atomic mass at the bottom-center.

Table 2.3 Molecular Weight of Ammonia (NH_3)

Element	No. Atoms × Atomic Weight	Weight of Element	% Weight of Molecule
Hydrogen	3 × 1	3	18
Nitrogen	1 × 14	14	82

With respect to nitrification, ammonia is one of two major reduced forms of nitrogen found in domestic wastewater. Ammonium is the other form. In these two reduced forms of nitrogen only hydrogen atoms are bonded to the nitrogen.

Ammonia is a weak base. It combines with acids such as muriatic or hydrochloric acid (HCl) to form a salt (Equations 2.1). The reaction for ammonia and hydrochloric acid results in the production of ammonium chloride (NH_4Cl).

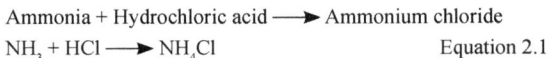

Ammonia + Hydrochloric acid $\longrightarrow$ Ammonium chloride

$NH_3 + HCl \longrightarrow NH_4Cl$ Equation 2.1

In domestic wastewater ammonia is a primary byproduct of the hydrolysis of urea and the degradation of proteins. The toxicity of ammonia is dependent upon temperature and pH. However, pH has a greater role in determining ammonia toxicity. With increasing pH ammonia becomes toxic because it forms caustic ammonium hydroxide (**NH_4OH**) when it dissolves in water (moist linings of the respiratory tract in humans).

Ammonia exists as a liquid at very low temperature and high pH. This form of ammonia is called anhydrous (water-free) ammonia. Ammonia is soluble in water, and an aqueous solution of ammonia is in the form of ammonium hydroxide (Equations 2.2).

Ammonia + Water $\longleftrightarrow$ Ammonium hydroxide $\longleftrightarrow$ Ammonium + Hydroxyl ion

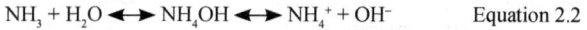

$NH_3 + H_2O \longleftrightarrow NH_4OH \longleftrightarrow NH_4^+ + OH^-$ Equation 2.2

Ammonia serves no useful role in the activated sludge process. Ammonia is toxic to bacteria, protozoa, rotifers, and soil nematodes (Figure 2.2). Ammonium is the primary nitrogen nutrient and the energy substrate for ammonium-oxidizing bacteria (AOB).

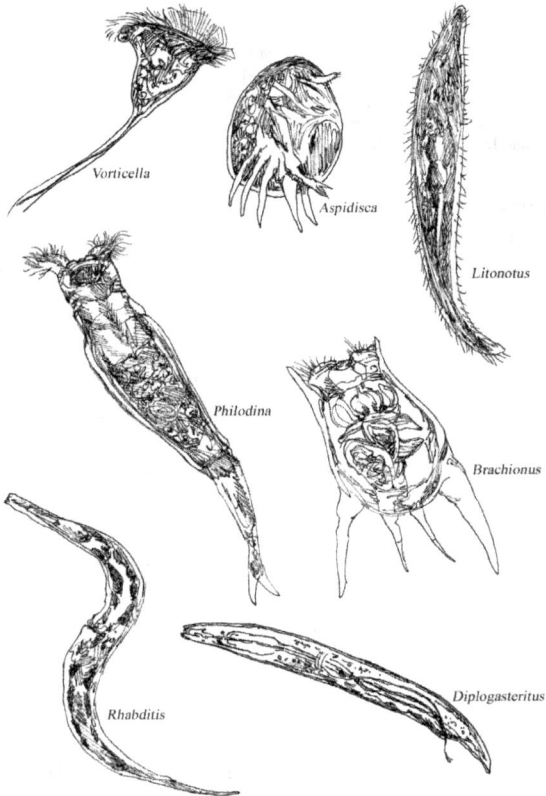

Figure 2.2 Commonly observed protozoa, rotifers, and soil nematodes in the activated sludge process: There are a large number and a large diversity of microscopic organisms or higher life forms (soil and water organisms) that enter the activated sludge process through inflow and infiltration (I/I). These organisms include ciliated protozoa, rotifers, and soil nematodes that perform several significant roles in the stabilization of wastewater in the activated

sludge process. Roles performed include cropping action or the consumption of dispersed bacterial cells and coating action or the removal of fine solids through their adsorption to floc particles. Commonly observed ciliated protozoa include *Vorticella* (stalk ciliate), *Aspidisca* (crawling ciliate), and *Litonotus* (free-swimming ciliate). Commonly observed rotifers include free-swimming *Brachionus* and crawling *Philodina*. Commonly observed soil nematodes (worms) include *Rhabditis* and *Diplogasteritus*.

Ammonium

Ammonium or ionized ammonia (NH_4^+) also is a form of reduced nitrogen. The molecular weight of ammonium is 18 g/mol (Table 2.4). Nitrogen makes up 78% of the weight of ammonium. Therefore, an analytical test value of 1 mg/L of NH_4^+ is the same as 0.78 mg/L of NH_4^+-N.

Table 2.4 Molecular Weight of Ammonium (NH_4^+)

Element	No. Atoms × Atomic Weight	Weight of Element	% Weight of Molecule
Hydrogen	4 × 1	4	22
Nitrogen	1 × 14	14	78

Ammonium is a weak acid and is found in a variety of salts including ammonium carbonate ((NH_4)$_2CO_3$) and ammonium nitrate (NH_4NO_3). The hydrogen atoms in ammonium can be substituted with an organic group to form substituted ammonium such as methylamine (CH_3NH_2), dimethylamine ((CH_3)$_2NH$), and trimethylamine ((CH_3)$_3N$) (Figure 2.3). Trimethylamine is one of many odorous compounds produced under an anaerobic condition. Quaternary ammonium has four substituted groups bonded to nitrogen (Figure 2.4). Quaternary ammonium or quats are referred to as quaternary ammonium compounds.

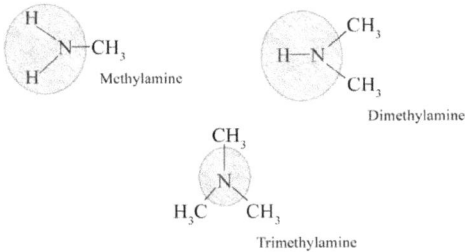

Figure 2.3 Methylamine, dimethylamine, and trimethylamine: Ammonium (NH_4^+) has four hydrogen atoms. Each hydrogen atom may be substituted by another atom or molecule such as the methyl group (-CH_3). If substitution occurs the ammonium may be converted, with one methyl group to methylamine, two methyl groups to dimethylamine, or three methyl groups to trimethylamine.

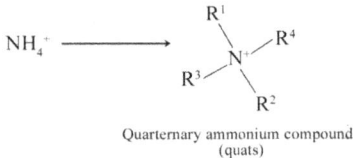

Quarternary ammonium compound
(quats)

Figure 2.4 Substituted ammonium compounds: When all four hydrogen atoms on ammonium (NH_4^+) are replaced (substituted) for other elements or molecules, ammonium is referred to as a quaternary ammonium compound. The compounds formed on the nitrogen atom are referred to as quaternary ammonium compounds. These compounds are highly toxic.

In domestic wastewater ammonium and ammonia are products of cellular metabolism. In municipal wastewater ammonium is a product of cellular metabolism and industrial discharges (Table 2.5). A major source of ammonium is the presence of urea (NH_2CONH_2), which is found in urine. When discharged to the sewer system urea undergoes hydrolysis. Hydrolysis occurs when hydrolytic bacteria add water to urea to split urea into ammonium and carbon dioxide (CO_2) (Equations 2.3) (Figure 2.5). If the influent temperature is 15 to 30°C, and the pH is 6.5 to 8.0, ammonium is the dominant form of reduced nitrogen.

Table 2.5 Industrial Discharges Containing Significant Quantities of Ammonium

| Automotive facilities |
| Chemical manufacturing |
| Coal gasification |
| Fertilizer manufacturing |
| Food-processing facilities |
| Landfill leachate |
| Livestock maintenance |
| Ordnance sites |
| Meat processing |
| Metal industries |
| Petrochemical |
| Refineries |
| Steel manufacturing |
| Tanneries |

Urea + Water —— Hydrolytic bacteria ➙ Ammonia* + Carbon dioxide

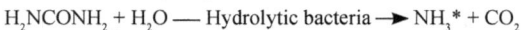

$H_2NCONH_2 + H_2O$ —— Hydrolytic bacteria ➙ NH_3* + CO_2

Equation 2.3

*at near neutral pH, ammonia is converted to ammonium

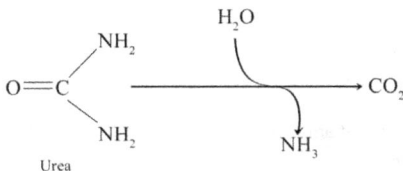

Urea

Figure 2.5 Hydrolysis of urea: Urea (NH_2COH_2N) is a major component of urine. It is also a highly soluble simplistic compound that undergoes hydrolysis in the sewer system resulting in the release of its amine groups (-NH_2)

and the formation of ammonia. Hydrolysis of urea occurs when hydrolytic bacteria add water to urea and split ("lysis") urea into amine groups and carbon dioxide (CO_2).

Ammonium serves two beneficial biological roles. First, it is the primary nutrient for nitrogen that is used for cellular growth (sludge production). Second, it is the energy substrate for ammonium-oxidizing bacteria (AOB).

Ammonia vs. ammonium

Ammonia (NH_3) is un-ionized, while ammonium (NH_4^+) is ionized. The quantity or percent of each in a treatment tank is determined by temperature and pH. However, the quantity of each is more dependent on pH. Ammonia is largely predominant at neutral or slightly alkaline wastewater. At pH 7.4 and temperature of 20°C the ammonium-to-ammonia concentration is approximately 100:1. At pH 8, ammonia is 10% of the reduced nitrogen present, and at pH 9, ammonia is 50% of the reduced nitrogen present. At pH 10, nearly all reduced nitrogen is in the ammonia form. At pH $\geq$ 10, ammonia can be stripped to the atmosphere. Within the pH range of 6.5 to 8.5 at most activated sludge processes, ammonium is the predominant form of reduced nitrogen.

The relationship between ammonia, ammonium, and pH is presented in Equation 2.4. When the pH is low, the reaction is driven to the right, and when the pH is high, the reaction is driven to the left. Some analyses report total ammonia (TA), which is the sum of ammonia (NH_3) and ammonium (NH_4^+), while some analyses report total ammonia-nitrogen, which is the sum of ammonia-nitrogen (NH_3-N) and ammonium-nitrogen (NH_{4+}-N).

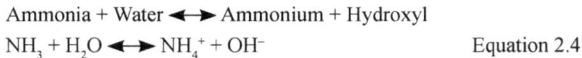

Ammonia + Water $\longleftrightarrow$ Ammonium + Hydroxyl

$NH_3 + H_2O \longleftrightarrow NH_4^+ + OH^-$ Equation 2.4

Amines

Amines are organic compounds that contain nitrogen. The term "organic nitrogen" is used to describe a nitrogenous compound that had its origin in a living cell. Many organic nitrogen compounds contain the amine group (-NH_2). Amino acids, the compounds that are used to build proteins, have the amine group (Figure 2.6).

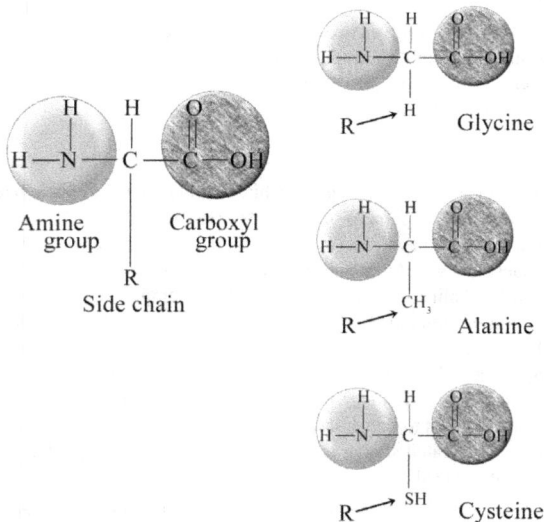

Figure 2.6 Basic structure of amino acids and examples of amino acids: Amino acids are the building blocks for proteins. All amino acids contain an acidic or carboxyl group (-COOH) and an amine group (NH_2) (upper left corner). On a non-carboxyl carbon a radical group (R) may be substituted for a hydrogen atom. Simplistic amino acids with substituted hydrogen atoms are glycine, alanine, and cysteine. Cysteine is one of only two amino acids that contain sulfur (S).

Amines are also derivatives of ammonia that have one or more hydrogen atoms replaced by an organic group (Figure 2.7). When organic nitrogen is degraded, the amine group is released and depending on the pH of the reactor, it is converted to ammonia or ammonium (Figure 2.8).

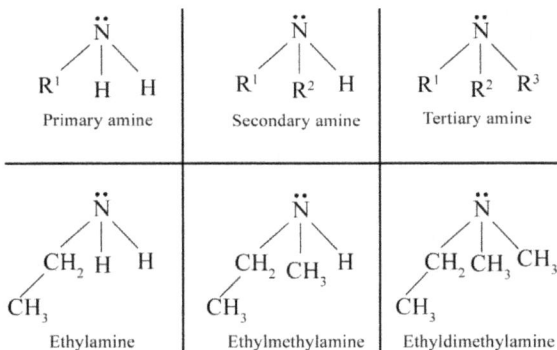

Figure 2.7 Examples of amines having substituted hydrogen atoms: The nitrogen in an amine group may have one, two, or three of its hydrogen atoms substituted with another element or compound. If one hydrogen atom is substituted, the amine is referred to as a primary amine such as ethylamine. If two hydrogen atoms are substituted, the amine is referred to as a secondary amine such as ethylmethylamine, while the substitution of three hydrogen atoms results in a tertiary amine such as ethyldimethylamine.

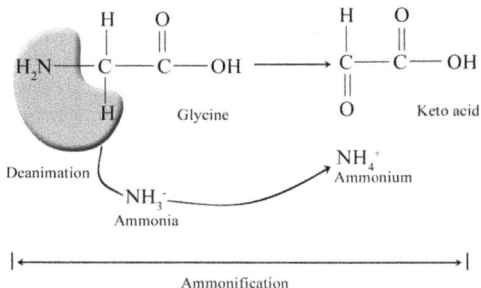

Figure 2.8 Deamination: When organic nitrogen compounds like amino acids and proteins are degraded, the amine ($-NH_2$) is release from the compound through deamination. Deamination is the cleavage of the amine group and an

adjacent hydrogen atom from the compound. When cleaved, the amine group and hydrogen atom form ammonia. If the pH of the bulk solution is < 9.0, ammonia is converted to ammonium.

Degradation of organic nitrogen and production of ammonia or ammonium is referred to as ammonification. The removal of the amine group is referred to as deamination.

Nitrite and nitrate

Nitrite and nitrate are two oxidized forms of nitrogen. They are produced in an activated sludge process that nitrifies. Nitrite is produced when ammonium-oxidizing bacteria (AOB) add oxygen to ammonium. Nitrite is an unstable and usually a short-lived nitrogenous compound that under favorable operational conditions is quickly oxidized to nitrate by nitrite-oxidizing bacteria (NOB). However, depressed temperature and limiting factors can prevent the oxidation of nitrite to nitrate. This permits the accumulation of nitrite in the activated sludge process. Nitrite is highly motile in wastewater and is highly toxic to aquatic life at low concentrations. Only a small number of bacteria can use nitrite as a nitrogen nutrient in the absence of ammonium.

Although toxic at high concentrations, nitrate serves as a nitrogen nutrient in the absence of ammonium. When used as a nitrogen nutrient this event is referred to as assimilatory nitrate reduction, because nitrogen is incorporated into cellular material. When used as a carrier molecule to remove electrons from degrading compounds in the cell, this event is referred to as dissimilatory nitrate reduction, because nitrogen is not incorporated into cellular material. Nitrogen leaves the cell. High concentrations (> 20 mg/L) of nitrate may result in the failure of whole effluent toxicity testing (WETT).

Nitrite

Nitrite (NO_2^-) is a compound of nitrogen and oxygen. The molecular weight of nitrite is 46 g/mol (Table 2.6). Nitrogen makes up 30% of the weight of nitrite. Therefore, an analytical test value of 1 mg/L of NO_2^- is the same as 0.3 mg/L of NO_2^--N.

Table 2.6 Molecular Weight of Nitrite (NO_2^-)

Element	No. Atoms × Atomic Weight	Weight of Element	% Weight of Molecule
Nitrogen	1 × 14	14	30
Oxygen	2 × 16	32	70

Nitrite is one of two oxidized forms of nitrogen found in an aerobic reactor where nitrification is occurring. Nitrate is the other form. In these two oxidized forms of nitrogen only oxygen atoms are bonded to the nitrogen.

Nitrite can be oxidized to nitrate or reduced to insoluble nitrogen-containing gases, nitrous oxide (N_2O) or molecular nitrogen (N_2). The oxidation of nitrite to nitrate is nitrification. The reduction of nitrate to nitrite is denitrification.

In wastewater, nitrite can form nitrous acid (HNO_2) during nitrification (Equations 2.5). Nitrous acid destroys alkalinity, is responsible for the occurrence of the chlorine sponge, and is toxic to aquatic life.

Nitrite + Proton ⬌ Nitrous acid

$$NO_2^- + H^+ \longleftrightarrow HNO_2 \hspace{2cm} \text{Equation 2.5}$$

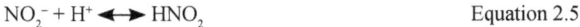

In an aerobic reactor and under favorable operational conditions ammonium is rapidly oxidized to nitrite (Equations 2.6). The oxidation of ammonium is the first step or reaction in the two-step nitrification process. Ammonium-oxidizing bacteria (AOB) perform the oxidation of ammonium to nitrite. Ammonium is the energy substrate for AOB.

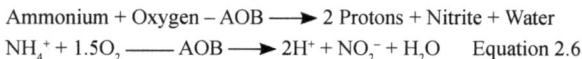

Ammonium + Oxygen – AOB $\longrightarrow$ 2 Protons + Nitrite + Water

$$NH_4^+ + 1.5O_2 \text{ ——— AOB} \longrightarrow 2H^+ + NO_2^- + H_2O \qquad \text{Equation 2.6}$$

Nitrate

Nitrite (NO_3^-) is a compound of nitrogen and oxygen. The molecular weight of nitrate is 62 g/mol (Table 2.7). Nitrogen makes up 23% of the weight of this compound. Therefore, an analytical test value of 1 mg/L of NO_3^- is the same as 0.23 mg/L of NO_3^--N.

Table 2.7 Molecular Weight of Nitrate (NO_3^-)

Element	No. Atoms × Atomic Weight	Weight of Element	% Weight of Molecule
Nitrogen	1 × 14	14	23
Oxygen	3 × 16	48	77

Nitrate is found in an aerobic reactor where nitrification occurs. In nitrate the nitrogen is in its highest oxidation state. The nitrogen atom cannot be oxidized further, but nitrate can be reduced to nitrite.

The oxidation of nitrite produces nitrate. This is the second step of the two-step nitrification process (Equation 2.7). Nitrite-oxidizing bacteria (NOB) perform the oxidation of nitrite to nitrate. Nitrite is the energy substrate for nitrite-oxidizing bacteria.

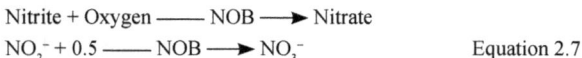

Nitrite + Oxygen ——— NOB $\longrightarrow$ Nitrate

$$NO_2^- + 0.5 \text{ ——— NOB} \longrightarrow NO_3^- \qquad \text{Equation 2.7}$$

Nearly all nitrate salts such as ammonium nitrate (NH_4NO_3) and potassium nitrate (KNO_3) are soluble in wastewater. Nitrate salts that are commonly used to prevent or control odor production in reactors or the sewer system are sodium nitrate ($NaNO_3$) and calcium nitrate ($Ca(NO_3)_2$). Nitrate as well as nitrite can be found in some industrial wastewaters (Table 2.8).

Table 2.8 Industrial Wastewaters That Contain Nitrate (NO_3^-) or Nitrite (NO_2^-)

Industrial Wastewater	Nitrite Present	Nitrate Present
Corrosion inhibitors	X	
Leachate (biological pretreatment)	X	X
Meat processing (flavor additives)		X
Meat processing (preservatives)	X	
Meat processing (biological pretreatment)	X	
Steel mill	X	X

Groups of nitrogenous compounds

In order to determine the groups of nitrogenous compounds in wastewater, four chemical tests must be performed. The tests include: 1) ammonia or ammonium, 2) nitrite, 3) nitrate, and 4) total Kjeldahl nitrogen (TKN). For process control separate nitrite and nitrate tests must be performed. A combined test for nitrite and nitrate may be acceptable to determine a total nitrogen value, but this test provides little information for process control.

The TKN test is used to determine the quantity of ammonium that can be released from organic nitrogen compounds through deamination and ammonification. Therefore, the TKN value minus the ammonium value from a separate test of the same sample equals the quantity of ammonium that can be released from organic nitrogen (Equation 2.8). The TKN value is either greater than or equal to the ammonium value.

TKN – ammonium = ammonium from organic nitrogen

Equation 2.8

Data from all four tests can be used to determine the following nitrogenous groups of compounds (Figure 2.9): 1) total nitrogen (TN), 2) total inorganic nitrogen (TIN), and 3) total organic nitrogen. Total nitrogen consists of TKN, nitrite, and nitrate. Total inorganic nitrogen consists of ammonium, nitrite, and nitrate, and organic nitrogen is the TKN minus the ammonium. Except for TKN, the reported values for the different analytical methods may be listed as:

- ammonia or ammonium (NH_3 or NH_4^+)
- ammonia-nitrogen or ammonium-nitrogen (NH_3-N or NH_4^+-N)
- nitrite or nitrite-nitrogen (NO_2^- or NO_2^--N)
- nitrate or nitrate-nitrogen (NO_3^- or NO_3^--N)

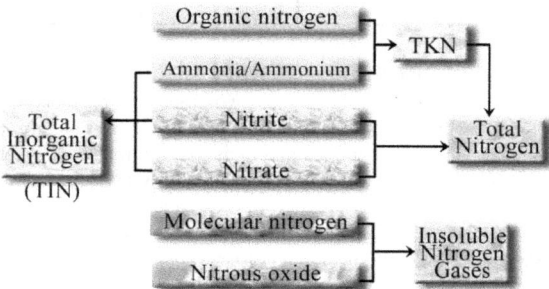

Figure 2.9 Groups of nitrogenous compounds: There are seven nitrogenous compounds of concern to wastewater treatment plant operators. These compounds are organic nitrogen, ammonia, ammonium, nitrite, nitrate, molecular nitrogen, and nitrous oxide. These compounds make up four groups of nitrogenous compounds of interest to wastewater treatment plant operators. These groups are total inorganic nitrogen (ammonia/ammonium, nitrite, nitrate), total Kjeldahl nitrogen (TKN) (organic nitrogen and ammonia/ammonium), total nitrogen (TN) (TKN, nitrite, and nitrate), and insoluble nitrogen gases (molecular nitrogen and nitrous oxide).

Based on the percent composition of nitrogen in each compound, the amount of nitrogen present is different for each respective pair of compounds (Table 2.9).

Table 2.9 Quantity of Nitrogen in 1 mg/L in Reported Test Values

Test	Quantity of Nitrogen Present in each mg/L
NH_3	0.8
NH_3-N	1.0
NH_4^+	0.8
NH_4^+-N	1.0
NO_2^-	0.3
NO_2^--N	1.0
NO_3^-	0.2
NO_3^--N	1.0

Hydrolysis of organic nitrogen compounds can be incomplete, if the compounds are recalcitrant. Also, some organic nitrogen compounds such as the moderately toxic acetonitrile (CH_3CN) are not measured by the TKN test.

Phosphorus

Phosphorus is found in wastewater in quantities greater than those needed by aquatic plants for growth. Influent phosphorus at municipal wastewater treatment plants usually is 10 to 20 mg/L and exists in inorganic and organic forms. Effluent phosphorus from municipal wastewater treatment plants typically is found in the orthophosphate form.

Phosphorus occurs mostly as phosphate (PO_4^{3-}). The major phosphorus compounds include: 1) phosphates, 2) condensed phosphates, and 3) organic phosphorus (Table 2.10). Common organic phosphorous compounds found in cellular waste include phytic acid, nucleic acids, and phospholipids. When these compounds are degraded by bacteria, orthophosphate is released.

Table 2.10 Major Phosphorous Compounds

Compounds	Source
Orthophosphates (reactive phosphorus) $H_2PO_4^-$, HPO_4^{2-}, PO_4^{3-}	Main constituent in fertilizers used for agriculture and residential purposes; found in surface water and groundwater
Condensed (inorganic phosphates)	Phosphorous compounds that contain salts and/or metals such as calcium, potassium, and sodium; used in industry and as food additives
Organic phosphorus	Formed primarily by biological processes; found in human waste and food residues

Groups of phosphorous compounds

Orthophosphate

Phosphoric acid (H_3PO_4) is also known as orthophosphoric acid. In water the acid dissociates and depending on the pH of the water it exists as three forms of orthophosphate: phosphate (PO_4^{3-}), monohydrogen phosphate (HPO_4^{2-}), and dihydrogen phosphate ($H_2PO_4^-$). These forms of phosphoric acid are commonly known as orthophosphates. Monohydrogen phosphate and dihydrogen phosphate are found in activated sludge processes that operate at neutral or a near neutral range of pH values (6.8 to 7.2). Monohydrogen phosphate is dominant at pH values > 7, while dihydrogen phosphate is dominant at pH values < 7.

Because the three forms of orthophosphate are nearly identical in composition and structure, one form may be used for operational calculations. The molecular weight of monohydrogen phosphate is 96 mol/g (Table 2.11). Phosphorus makes up 32% of the weight of this compound. Therefore, an analytical test value of 1 mg/L of orthophosphate is the same as 0.32 mg/L of HPO_4^{2-}-P.

Table 2.11 Molecular Weight of Orthophosphate (HPO_4^{2-})

Element	No. Atoms × Atomic Weight	Weight of Element	% Weight of Molecule
Hydrogen	1 × 1	1	1
Phosphorus	1 × 31	31	32
Oxygen	4 × 16	64	67

Orthophosphate is the only form of phosphorus that can be used by bacteria as a nutrient source for phosphorus. Because orthophosphate also is the only form of phosphorus that reacts with a coagulant in order to be precipitated from the wastewater, it is referred to as "reactive" phosphorus (Table 2.12).

Table 2.12 Coagulants Used to Precipitate Orthophosphate

Group	Name	Formula
Aluminum salts	Aluminum chloride	$AlCl_3$
	Aluminum chlorohydrate Poly-aluminum chloride (PQC)	$Al_nCl_{(3n-m)}(OH)_m$
	Aluminum sulfate (alum)	$Al_2(SO_4)_3$
	Sodium aluminate	$Na_2Al_2O_4$
Iron salts	Ferric chloride	$FeCl_3$
	Ferrous chloride	$FeCl_2$
	Ferrous sulfate	$FeSO_4$
Calcium salts	Hydrated lime	$Ca(OH)_2$
	Limestone	$CaCO_3$

Total phosphorus

In addition to orthophosphate there are two groups of phosphorous compounds, condensed phosphates or polyphosphates and organic phosphorus or organophosphate (Figure 2.10). Condensed phosphates and organic phosphorus do not serve as a nutrient source for phosphorus. They must be degraded or hydrolyzed in order to release orthophosphate (Figure 2.11). Also, a coagulant cannot precipitate phosphorus in these compounds, unless the compounds are degraded or hydrolyzed to release orthophosphate. If the compounds cannot be degraded or hydrolyzed, they are adsorbed to floc particles or solids, or they are lost from the plant in the effluent. Bacteria synthesize organic phosphorous compounds as well as organic nitrogen compounds. Organic phosphorous compounds are synthesized from polyphosphates. Because bacteria contain organic phosphorus and organic nitrogen, a loss of solids in the effluent represents an increase in TP and TN.

Since a bacterial cell may contain approximately 3% phosphorus on a dry weight basis and approximately 15% nitrogen on a dry weight basis, a loss of 30 mg/L solids in the effluent could represent a loss of approximately 1.0 mg/L total phosphorus and 4.5 mg/L total nitrogen. These effluent values for total phosphorus and total nitrogen may exceed the discharge limit for phosphorus and nitrogen. Phosphorus is also found in detritus and enters the activated sludge process through inflow and infiltration (I/I).

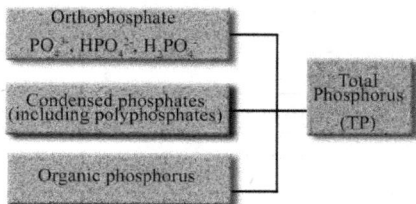

Figure 2.10 Groups of phosphorous compounds: Three are three groups of phosphorous compounds that make up total phosphorus (TP). These compounds consist of orthophosphate (PO_4^{3-}, HPO_4^{2-}, and $H_2PO_4^{-}$), condensed phosphate, and organic phosphorus.

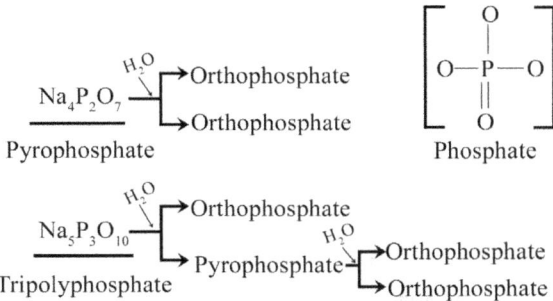

Figure 2.11 Hydrolysis of polyphosphate: Polyphosphate is a long-chain polymer of phosphate or chain of mers of phosphate. Polyphosphate undergoes hydrolysis in the activated sludge process. The chain of mers is hydrolyzed from tripolyphosphate (3 phosphate unit mers) to orthophosphate (1 phosphate unit) and pyrophosphate (2 phosphate mers). Pyrophosphate is hydrolyzed to orthophosphate.

Condensed phosphates or polyphosphates are water-soluble acids (Figure 2.12). Polyphosphates are used in baking powder, paint strippers, soft drinks, and toothpaste. They may be added to municipal water systems to prevent corrosion in pipes. They are polymers of phosphate (Figure 2.13). If sufficient time is provided in the activated sludge process, polyphosphates can be hydrolyzed to orthophosphate. Based on the percent composition of phosphorus in each compound, the amount of the phosphorus present is different for orthophosphate and orthophosphate-phosphorus (Table 2.13).

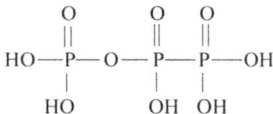

Triphosphoric acid

Figure 2.12 Polyphosphate: Polyphosphates such as triphosphoric acid consist of numerous phosphate units bonded together. When polyphosphate is hydrolyzed, phosphate is released.

$$O = \overset{\displaystyle O^-}{\underset{\displaystyle O^-}{P}} - O^-$$

$$(PO_4^{3-})$$

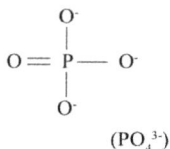

Figure 2.13 Phosphate: Phosphate or orthophosphate is the only form of phosphorus that can be used by bacteria as a nutrient source for phosphorus. It also is the only form of phosphorus that reacts with a coagulant for chemical precipitation of phosphorus. Because phosphate or orthophosphate reacts with a coagulant, it is called "reactive" phosphorus.

Table 2.13 Quantity of Phosphorus in 1 mg/L in Reported Test Values

Test	Quantity of Phosphorus Present in each mg/L
HPO_4^{2-} or $H_2PO_4^-$	0.3
HPO_4^{2-}-P or $H_2PO_4^-$-P	1.0

Orthophosphate is analyzed directly on an unpreserved wastewater sample within 48 hours of sampling. Total phosphorus is analyzed from a sulfuric acid (H_2SO_4) preserved wastewater sample within 28 days of sampling. Total phosphorus is determined by converting all forms of phosphorus to orthophosphate using acid hydrolysis.

Chapter 3

Nitrification: The Basics

Nitrification is the oxidation of ammonium (NH_4^+) to nitrite (NO_2^-) and/or the oxidation of nitrite to nitrate (NO_3^-) (Equation 3.1). In soil, living nitrifying bacteria perform the two-step conversion of ammonium to nitrate. Ammonium-oxidizing bacteria (AOB) perform the first step, the oxidation of ammonium to nitrite (Equations 3.2). The most commonly cited genus of AOB is *Nitrosomonas*. Nitrite-oxidizing bacteria (NOB) perform the second step, the oxidation of nitrite to nitrate (Equations 3.3). The most commonly cited genus of NOB is *Nitrobacter*. The rate of each step is determined mostly by temperature.

Ammonium + Oxygen ——— AOB ⟶ NOB ⟶ Nitrate + Protons + Water

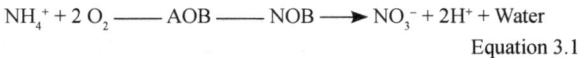

$$NH_4^+ + 2\,O_2 \text{——— AOB ——— NOB} \longrightarrow NO_3^- + 2H^+ + \text{Water}$$

Equation 3.1

Ammonium + Oxygen ——— AOB ⟶ Protons + Nitrite + Water

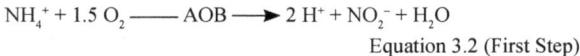

$$NH_4^+ + 1.5\,O_2 \text{——— AOB} \longrightarrow 2\,H^+ + NO_2^- + H_2O$$

Equation 3.2 (First Step)

Nitrite + Oxygen ——— NOB ⟶ Nitrate

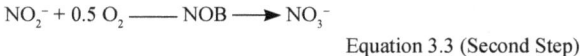

$$NO_2^- + 0.5\,O_2 \text{——— NOB} \longrightarrow NO_3^-$$

Equation 3.3 (Second Step)

AOB oxidize ammonium in order to obtain energy. Ammonium is their energy substrate. NOB oxidize nitrite in order to obtain energy. Nitrite is their energy substrate. However, very little energy is obtained by the oxidation of ammonium and nitrite. With very little energy for cellular activity, generation time for each group of nitrifying bacteria is relatively long. During these energy-yielding reactions some of the ammonium is not oxidized to nitrite but is assimilated into new cellular material (sludge).

Nitrifying bacteria use only inorganic compounds, ammonium, and nitrite as their energy substrate (Table 3.1). These two nitrogenous compounds make up the nitrogenous BOD (nBOD). Nitrifying bacteria do not use organic or carbonaceous compounds (cBOD) as energy substrates, nor do they use carbonaceous compounds as a source of carbon for growth. Nitrifying bacteria use alkalinity and prefer bicarbonate alkalinity (HCO_3^-) as their carbon source. If an adequate quantity of alkalinity is not available, growth of nitrifying bacteria does not occur. If growth does not occur then nitrification stops.

Table 3.1 Sources of Ammonium, Nitrite, and Nitrate

Waste/Wastewater	Ammonium (NH_4^+)	Nitrite (NO_2^-)	Nitrate (NO_3^-)
Domestic wastewater	X		
Fertilizer manufacturing	X		
Inhibitors of corrosion		X	
Landfill leachate	X		
Leachate, biologically pretreated		X	X
Livestock maintenance	X		
Meat processing	X		
Meat processing, biologically pretreated		X	X
Meat processing, flavor additives			X
Meat processing, preservation		X	
Petrochemical	X		
Steel		X	X

Nitrite and nitrate typically are not found in the influent to municipal activated sludge processes. It is rare to find nitrite concentrations > 1 mg/L in water or wastewater. If nitrite or nitrate is found in the sewer system, it is due to industrial discharge. There are four factors that are responsible for the inhibition of nitrification in the sewer system (Figure 3.1):

- the presence of a small population of nitrifying bacteria
- short hydraulic retention time (HRT)
- low dissolved oxygen level
- presence of inhibitory and toxic compounds including hydrogen sulfide (H_2S) and recognizable, soluble cBOD

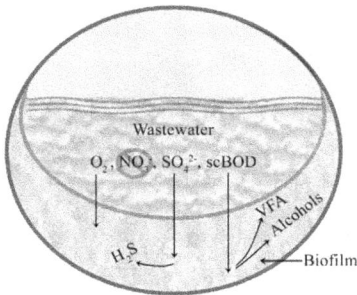

Figure 3.1 Inhibition of nitrification in the sewer system: There are several reasons for the inhibition of nitrification in the sewer system. These include low dissolved oxygen, low population of nitrifying bacteria, short retention time, and the presence of inhibitory and toxic compounds. The inhibitory and toxic compounds are produced in the sewer system by the bacterial activity in the sediment and biofilm in the sewer system. In the absence of dissolved oxygen or presence of residual dissolved oxygen and absence of nitrate, bacteria in the sediment and biofilm use sulfate (SO_4^{2-}) to degrade soluble cBOD. When sulfate is used, hydrogen sulfide (H_2S) is used. Hydrogen sulfide is toxic to nitrifying bacteria. Soluble cBOD also is degraded through fermentation. The fermentative process releases volatile fatty acids and a mixture of alcohols. Many of these fermentative products are inhibitory to nitrifying bacteria.

In addition to the presence of an abundant and active population of nitrifying bacteria and favorable wastewater temperature, there are several operational measures or limiting factors that influence the activity of nitrifying bacteria and consequently nitrification. These factors must be acceptable in order to achieve proper nitrification. The limiting factors that must be addressed are:

- alkalinity and pH
- dissolved oxygen concentration
- hydraulic retention time (HRT)
- inhibition and toxicity including substrate toxicity
- organic loading and cBOD-to-TKN ratio
- orthophosphate concentration
- population size of heterotrophic bacteria
- slug discharge of soluble cBOD

Nitrification can be complete or incomplete. Complete nitrification occurs when the mixed liquor effluent filtrate from an in-line aeration tank yields the following profile for ammonium-nitrogen, nitrite-nitrogen, and nitrate-nitrogen:

- Ammonium-nitrogen = ≤ 1 mg/L
- Nitrite-nitrogen = ≤ 1 mg/L
- Nitrate-nitrogen = ≤ 1 mg/L

If any limiting factor should adversely impact nitrification, then incomplete or partial nitrification occurs. Some forms of incomplete nitrification result in the production and accumulation of nitrite. The accumulation of nitrite can only be determined by nitrite testing alone. A combined nitrite/nitrate test does not determine if nitrite is present.

Nitrite accumulation

In some forms of incomplete nitrification nitrite is produced and accumulates. This can occur due to depressed temperature and/or a limiting factor. Nitrite is toxic and produces a significant chlorine demand that interferes with chlorine disinfection of the effluent and chlorination of the mixed liquor or return activated sludge (RAS) to control undesired filamentous organism growth. For every mg/L NO_2^--N present, 5 mg/L of chlorine are consumed. Nitrite is oxidized to nitrate by chlorine (Equations 3.4, 3.5, and 3.6). This large consumption of chlorine is known as the "chlorine sponge," "nitrite kick," and "nitrite lock."

$$Cl_2 + H_2O \longrightarrow HCl + HOCl$$

Chlorine gas dissolves in water and produces hydrochloric acid and hypochlorous acid.

Equation 3.4

$$HOCl \longleftrightarrow H^+ + OCl^-$$

Hypochlorous acid dissociates to form a proton and hypochlorite.

The degree of dissociation is dependent on pH and temperature.

Equation 3.5

$$OCl^- + NO_2^- \ ----- > NO_3^- + Cl^-$$

Hypochlorite reacts with nitrite and oxidizes it to nitrate with chloride as a byproduct.

Equation 3.6

The chlorine demand is often erratic, and the demand can be large enough to prevent the maintenance of proper chlorine residual. The inability to obtain proper chlorine residual can result in violations of the effluent discharge limit for an indicator organism such as *Escherichia coli* or indicator group such as fecal coliforms.

To correct for the "chlorine sponge" there are three operational measures: 1) the flow (MGD) and nitrite concentration should be monitored hourly; 2) the pounds of chlorine needed in addition to normal chlorine use or demand should be calculated; and 3) the chlorine feed should be adjusted. If the plant does not have an ammonia discharge limit or a total nitrogen discharge limit, an operator may add ammonia to the nitrite-laden wastewater. When ammonia is added, nitrite reacts with the ammonia and together they form chloramines. However, the best measure for correcting the chlorine sponge is to determine if cold temperature and/or a limiting factor is responsible for its development and compensate for cold temperature and/or correct the limiting factor.

Increase in ammonium in mixed liquor effluent filtrate

An increase in ammonium in the mixed liquor effluent filtrate can be due to operational conditions other than a loss of nitrification or the occurrence of incomplete nitrification. There conditions include:

- increase in ammonium in the mixed liquor influent
- increase in easily degradable organic-nitrogen in mixed liquor influent
- increase in hydraulic retention time in the aeration tank for deamination and ammonification to occur
- presence of excess polyacrylamide polymer used for dewatering operations
- increase in cBOD loading
- increase in soluble cBOD loading
- presence of nitrogenous compounds in recycle streams to the aeration tank

Recycle streams that typically contain a significant quantity of nitrogenous compounds and possibly phosphorous compounds that should be monitored for nitrogen are decant streams from aerobic and anaerobic digesters as well as the filtrate and centrate from belt filter press and centrifuge operations, respectively. If these recycle streams contain excess nitrogenous wastes such as ammonium or organic-nitrogen, the appropriate operations should be regulated with respect to the time of day when they are practiced in order to reduce the nitrogenous loading to the aeration tank or denitrification process. Additional recycle streams include: 1) return activated sludge (RAS), 2) thickener overflow, and 3) leachate from sand filters and reed beds.

Although much attention is given to the quantity of digester sludge to be dewatered and the compatibility of digester sludge with polymers, attention often is not given to the quantity and quality of nitrogenous compounds in digester decant and filtrate or centrate from dewatering operations. Compounds within these waste streams that are recycled to the activated sludge process may have a significant demand for dissolved oxygen for ammonification and nitrification and carbon for denitrification.

The aerobic digester is capable of nitrification (Figure 3.2). The cBOD loading to the digester is mostly colloidal and particulate in nature and degrades slowly. The cBOD loading to the digester does not have an immediate demand for oxygen and therefore provides an opportunity for nitrification. The solids inventory or biomass in the digester is typically more concentrated than the mixed liquor and the digester has a greater solids retention time (SRT) than the mixed liquor. Under these conditions nitrification occurs. Therefore, decant and sludge from the digester would contain relatively large quantities of nitrite and/or nitrate.

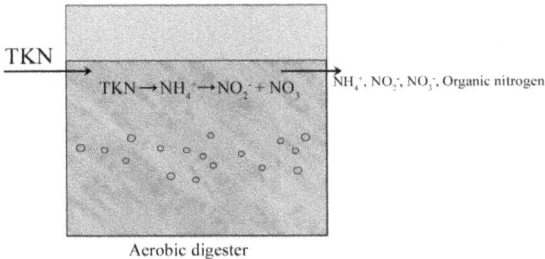

TKN

$TKN \rightarrow NH_4^+ \rightarrow NO_2^- + NO_3^-$ NH_4^+, NO_2^-, NO_3^-, Organic nitrogen

Aerobic digester

Figure 3.2 Aerobic digester and nitrogenous compounds: Colloidal cBOD, particulate BOD, and organic nitrogen compounds that are not degraded in the activated sludge process may be transferred to an aerobic digester. In the digester the organic nitrogen compounds undergo deamination and ammonification. The ammonium produced in the digester may be oxidized to nitrite and/or nitrate depending on operational conditions. Therefore, decant from the aerobic digester or centrate and filtrate from dewatering operations can contain relatively large quantities of nitrogen. The nitrogenous compounds include ammonium, nitrite, and nitrate as well as organic nitrogen in the form of solids (bacterial cells).

If the digester is not monitored for alkalinity, pH, nitrite, and nitrate, and a steady-state condition is not maintained for acceptable alkalinity and pH, the digester sludge may drop to pH values < 6.0. At low pH values toxicity can occur. It is therefore critical to monitor and maintain acceptable alkalinity and pH in the aerobic digester to prevent toxicity.

Due to the absence of dissolved oxygen, nitrification cannot occur in the anaerobic digester (Figure 3.3). Like the aerobic digester, the cBOD loading to the digester is mostly colloidal and particulate in nature, and the solids inventory or biomass is more concentrated than the mixed liquor. The solids retention time of the digester also is greater than the mixed liquor. Under these operational conditions ammonification occurs, resulting in a build-up of ammonium in the digester decant and sludge. If the ammonia concentration is > 1500 mg/L, ammonia toxicity can occur in the digester.

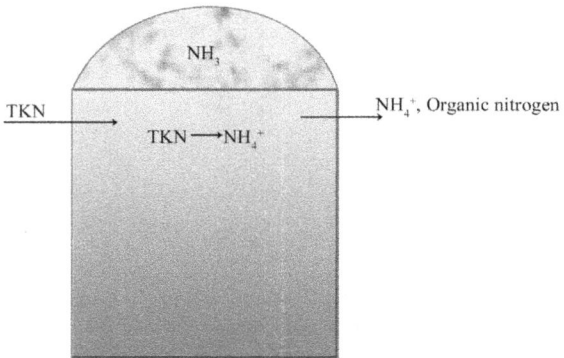

Figure 3.3 Anaerobic digester and nitrogenous compounds: Colloidal cBOD, particulate BOD, and organic nitrogen compounds that are not degraded in the activated sludge process may be transferred to an anaerobic digester. In the absence of free molecular oxygen, nitrification cannot occur. Therefore, deamination and ammonification of organic nitrogen compounds result in the production of ammonium. Therefore, decant from the anaerobic digester or centrate and filtrate from dewatering operations can contain relatively large

quantities of nitrogen. The nitrogenous compounds include ammonium and organic nitrogen in the form of solids (bacterial cells). With increasing pH in the digester the ammonium in converted to ammonia. At 1500 mg/L ammonia is toxic to methanogenic organisms in the digester.

Nitrification can occur in one-stage (single-stage) and two-stage processes (Figure 3.4). In the one-stage nitrification process cBOD removal and nitrification occur in the same aeration tank. In the two-stage nitrification process cBOD removal occurs in the first aeration tank and nitrification occurs in the second aeration tank. The two-stage nitrification process provides for longer retention time for nitrification and lowers the food-to-microorganism (F/M) ratio and cBOD-to-TKN ratio, which favors nitrification. Also, the first stage reduces or removes inhibitory and toxic wastes that impact nitrifying bacteria.

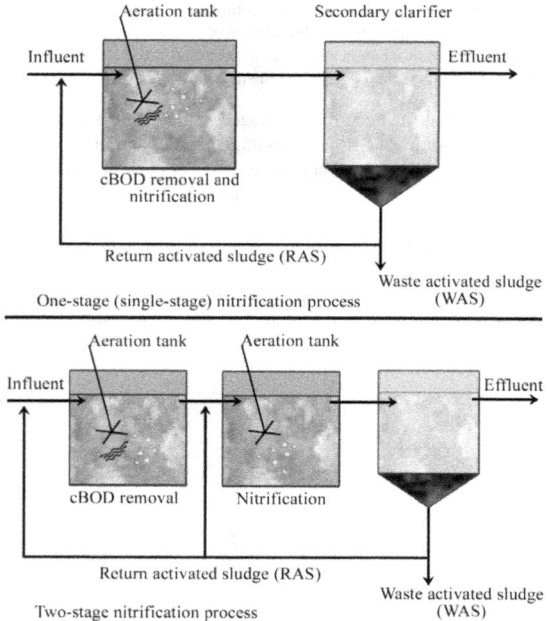

Figure 3.4 One-stage and two-stage nitrification processes: In one-stage nitrification processes the aeration tank or all aeration tanks degrade cBOD and nitrify. In two-stage nitrification processes the first aeration tank or series of aeration tanks remove cBOD, while the second aeration tank or series of aeration tanks nitrify. The two-stage nitrification system is better suited for operation in temperate regions of North America. It protects the nitrifying bacteria from inhibitory and toxic wastes, and reduces cBOD to a relatively low concentration even in cold wastewater temperatures, while the second stage of the two-stage nitrification processes can be adjusted to a pH that is favorable for nitrification. Also, the relative abundance of nitrifying bacteria can be increased in the second stage without an increase in cBOD-removing (heterotrophic) bacteria.

Demonstrating nitrification

A reduction in ammonium concentration across an aeration tank is not an indicator of nitrification. Although the concentration of ammonium is reduced through nitrification, it also is reduced through its assimilation into bacterial growth (sludge production) and air stripping as ammonia at high pH values. Therefore, the production of nitrite or nitrate must occur in the aeration tank to demonstrate nitrification. If the aeration tank nitrifies, then nitrite or nitrate must be present in a sample of mixed liquor effluent filtrate. However, if nitrification occurs then there will be strong indictors of nitrification based on differences in aeration tank influent and effluent values for alkalinity, pH, ammonium, nitrite/nitrate, and oxidation-reduction potential (ORP) (Figure 3.5).

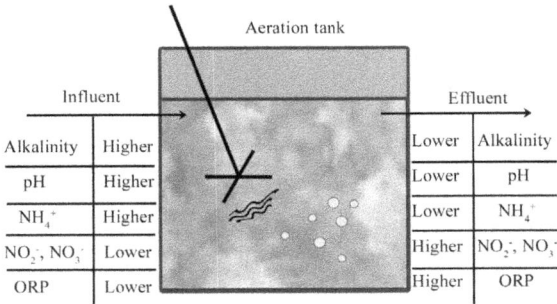

	Influent	Aeration tank	Effluent	
Alkalinity	Higher		Lower	Alkalinity
pH	Higher		Lower	pH
NH_4^+	Higher		Lower	NH_4^+
NO_2^-, NO_3^-	Lower		Higher	NO_2^-, NO_3^-
ORP	Lower		Higher	ORP

Figure 3.5 Indicators of nitrification across an aeration tank: If nitrification occurs in an aeration tank the following changes will occur with these specific parameters from influent to effluent of the aeration tank: 1) decrease in alkalinity as it is used as a carbon source for nitrifying bacteria and destroyed by the production of nitrous acid, 2) a decrease in pH as alkalinity is lost, 3) a decrease in ammonium as it is used as a nitrogen nutrient for bacterial growth and an energy substrate for nitrifying bacteria, 4) the production of nitrite and/or nitrate through nitrification, and 5) an increase in oxidation-reduction potential (ORP) as reduced wastes (cBOD and ammonium) are oxidized by bacteria using dissolved oxygen.

There are several indirect indicators of nitrification including the growth of duckweed (Figure 3.6) and filamentous algae in the secondary clarifier. Duckweed and algae use nitrate as a nitrogen nutrient. Therefore, nitrification must occur in the aeration tank in order for these organisms to proliferate. An increase in dissolved oxygen demand occurs after cBOD degradation has occurred. Nitrification occurs after cBOD degradation has occurred and requires a relatively large quantity of dissolved oxygen (approximately 4.6 lbs. oxygen per lb. ammonium oxidized to nitrate). Under optimal conditions for nitrification, numerous stalk ciliate protozoa in the genus *Epistylis* can be observed in a microscopic examination of mixed liquor (Figure 3.7). The protozoa proliferate under optimal conditions for nitrification, and they also nitrify.

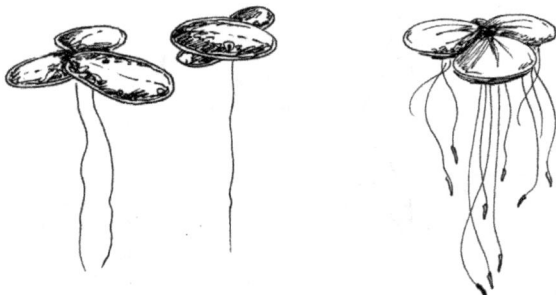

Lemna sp., common duckweed   *Spirodela* sp., giant duckweed

Figure 3.6 Duckweed: Duckweed is the smallest flower plant. It has white petals and floats on the surface of the secondary clarifier. Duckweed uses nitrate produced through nitrification in the aeration tank as its nitrogen nutrient. There are two genera of duckweed commonly observed on the surface of the secondary clarifier, *Lemna* and *Spirodela*.

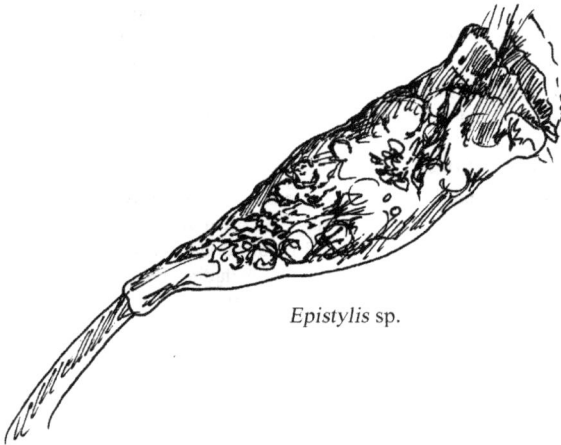

Epistylis sp.

Figure 3.7 *Epistylis:* The genus *Epistylis* consists of stalk ciliate protozoa that are found in large numbers in the activated sludge process when operational conditions are optimal for nitrification. Species of *Epistylis* are capable of limited heterotrophic nitrification.

Like the decrease in quantity of ammonium there are several events that also reduce the concentration of nitrite, nitrate, and organic nitrogen in an activated sludge process. Nitrite can be oxidized to nitrate not only biologically through nitrification but also chemically in the presence of hypochlorite (OCl^-). Although there are few genera of bacteria that can use nitrite as a nitrogen nutrient in the absence of ammonium, they are found in the activated sludge process and do use nitrite. However, the reduction of nitrite for use as a nitrogen nutrient is minor.

Nitrite and nitrate are reduced in concentration through denitrification. Also, nitrate is used and reduced in concentration by heterotrophic bacteria as a nitrogen nutrient for growth in the absence of ammonium.

Organic nitrogen can undergo deamination and ammonification with adequate retention time in the sewer system biofilm, aeration tank, aerobic digester, or anaerobic digester. Organic nitrogen in the colloidal and particulate forms is adsorbed to the surface of bacterial cells before deamination and ammonification occur.

Mass balance

A mass balance for nitrogen across a wastewater treatment process is complicated, time consuming, and often difficult to achieve. The mass balance must include all oxidized and reduced forms of nitrogen in numerous waste streams including recycle streams and nitrogenous gases released to the atmosphere.

Chapter 4

Nitrifying Bacteria

Many organisms are capable of performing nitrification, but only nitrifying bacteria are capable of performing sustained, high-rate nitrification. This ability is due to the in-folding of the cell membrane (Figure 4.1). The in-folding provides increased surface area for the placement of numerous nitrifying enzymes.

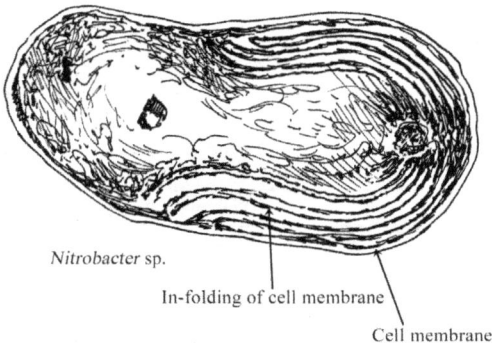

Figure 4.1 In-folding of cell membrane in nitrifying bacteria: Unlike some species of algae, fungi, and protozoa that are capable of limited or insignificant heterotrophic nitrification, nitrifying bacteria like *Nitrobacter* are capable of significant nitrification. This ability is due to the in-folding of the cell membrane. The in-folding provides increased surface area on the cell membrane for the placement of numerous nitrifying enzyme systems.

Nitrifying bacteria are strict aerobic, free-living soil and water organisms and enter the activated sludge process through inflow and infiltration (I/I). They are fragile organisms with a slow growth rate or generation time (three to six days) in the activated sludge process.

The most commonly referenced or popular nitrifying bacteria are *Nitrosomonas* and *Nitrobacter* (Table 4.1). *Nitrosomonas* oxidizes ammonium to nitrite. It is an ammonium-oxidizing bacterium (AOB). *Nitrosolobus*, *Nitrosouva*, and *Nitrosospira* are also ammonium-oxidizing bacteria. *Nitrobacter* oxidizes nitrite to nitrate. It is a nitrite-oxidizing bacterium (NOB). Other NOB include *Nitrospina* and *Nitrospiria*. Nitrifying bacteria are autotrophic organisms because they use carbon dioxide (CO_2) as their source of carbon for cellular growth.

Table 4.1 Characteristics of *Nitrosomonas* and *Nitrobacter*

Characteristic	*Nitrosomonas*	*Nitrobacter*
Carbon substrate	Inorganic (CO_2)	Inorganic (CO_2)
Energy substrate	Ammonium (NH_4^+)	Nitrite (NO_2^-)
Habitat	Soil and water	Soil and water
Oxygen requirement	Strict aerobic bacteria	Strict aerobic bacteria
pH growth range (laboratory)	5.8–8.5	6.5–8.5
pH growth range (activated sludge)	Typically 6.8 to 7.2	Typically 6.8 to 7.2
Generation time (laboratory)	8 to 36 hours	10 to 60 hours
Generation time (activated sludge)	3 to 6 days	3 to 6 days
Temperature growth range	5 to 30°C	5 to 40°C
Sludge yield (lbs. bacteria or sludge per pound of energy substrate oxidized)	0.04 to 0.13	0.02 to 0.07

Nitrifying bacteria are autotrophs, and cBOD degrading bacteria are heterotrophs (Table 4.2). There are two groups of nitrifying bacteria, ammonium-oxidizing bacteria (AOB) such as *Nitrosomonas* and nitrite-oxidizing bacteria (NOB) such as *Nitrobacter*. Unless nitrite is discharged to the activated sludge process, NOB is dependent upon AOB to provide its energy substrate nitrite (Figure 4.2). In order to nitrify in the activated sludge process, autotrophic nitrifying bacteria are dependent upon heterotrophic bacteria. Heterotrophic bacteria perform the following roles that promote nitrification:

- reduce soluble cBOD to < 15 mg/L
- form floc particles where nitrifying bacteria attach
- degrade organic nitrogen and produce ammonium
- remove inhibitory and toxic organic compounds

Table 4.2 Differences between Autotrophic Nitrifying Bacteria and cBOD-Degrading Heterotrophic Bacteria

Requirement or Activity	Autotrophic Nitrifying Bacteria	cBOD-Degrading Heterotrophic Bacteria
Carbon substrate	Inorganic compounds (alkalinity)	Organic compounds (cBOD)
Energy substrate	Ammonium and nitrite (nBOD)	Organic compounds (cBOD)
Generation time under optimal conditions	8–10 hours	20–30 minutes
Oxygen requirement	Yes, strict aerobic bacteria	Some, most are facultative anaerobic bacteria

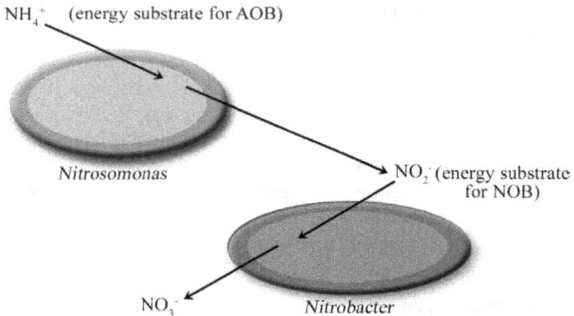

NH_4^+ (energy substrate for AOB)

Nitrosomonas

NO_2^- (energy substrate for NOB)

NO_3^- *Nitrobacter*

Figure 4.2 Dependency of NOB upon AOB for its energy substrate: Unless an industrial wastewater containing nitrite enters the activated sludge process, nitrite-oxidizing bacteria (NOB) are dependent upon ammonium-oxidizing bacteria (AOB) for their energy substrate for growth. AOB such as *Nitrosomonas* oxidize their energy substrate ammonium to nitrite and release the nitrite to the bulk solution. NOB such as *Nitrobacter* absorb nitrite from the bulk solution and use it as their energy substrate by oxidizing nitrite to nitrate. NOB release nitrate to the bulk solution.

By reducing soluble cBOD to < 15 mg/L heterotrophic bacteria provide nitrifying bacteria the opportunity to compete successfully for dissolved oxygen. Nitrifying bacteria make up approximately 10% of the bacterial biomass, while heterotrophic bacteria make up approximately 90% of the bacterial biomass. Most heterotrophic bacteria have higher metabolic rates than do nitrifying bacteria, and most heterotrophic bacteria have a shorter generation time than nitrifying bacteria. Therefore, during high BOD loading conditions when large quantities of dissolved oxygen are needed to degrade the BOD, heterotrophic bacteria can outcompete nitrifying bacteria for dissolved oxygen.

Increases and decreases in the population of nitrifying bacteria are strongly impacted by the cBOD-to-TKN ratio (Figure 4.3). Increasing the rate of nitrification can be achieved by increasing the population of nitrifying bacteria. By lowering the cBOD-to-

TKN ratio entering the aeration tank, the population size of heterotrophic bacteria is reduced without an accompanying decrease
in the population size for nitrifying bacteria. Thus the reduction in
cBOD-to-TKN ratio increases the percentage of nitrifying bacteria
in the bacterial community and promotes the growth of nitrifying
bacteria. To reduce the cBOD loading to the aeration tank, more
colloidal and particulate cBOD must be removed upstream of the
aeration tank. Operational measures available to decrease the cBOD
loading include placing more primary clarifiers in-line to provide
more settling time and adding a polymer or coagulant (metal salt) to
precipitate colloidal and particulate cBOD.

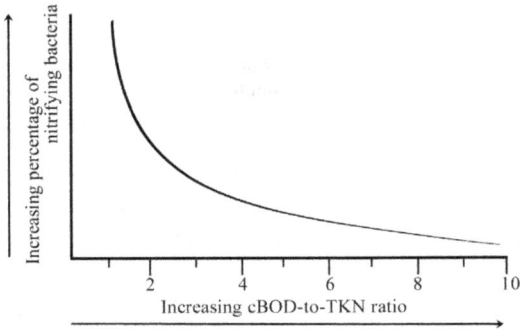

Figure 4.3 Relationship between fraction of nitrifying bacteria and changes
in cBOD-to-TKN ratio: By decreasing the quantity of cBOD entering the
aerobic reactor responsible for nitrification, the relative population size of
cBOD-removing or heterotrophic bacteria is reduced. This reduction in heterotrophic population increases the percentage of nitrifying bacteria in the
biomass. With less cBOD entering the reactor and a larger percentage of nitrifying bacteria in the biomass, nitrification is more easily achieved.

Nitrifying bacteria are poor floc formers. Therefore, they need to
be adsorbed to a surface in order to remain in the treatment process,
or they will be lost in the effluent. Heterotrophic bacteria form floc
particles to which nitrifying bacteria are adsorbed.

There are some short-chain, organic molecules (cBOD) that inhibit nitrifying bacteria. Heterotrophic bacteria degrade these organic molecules. In addition to the degradation of these organic molecules, heterotrophic bacteria may degrade larger organic compounds that are inhibitory or toxic, such as aromatic compounds like phenol (C_6H_5OH).

Number of nitrifying bacteria

An adequate population size of active nitrifying bacteria is necessary to achieve successful nitrification. The time required for nitrification is directly proportional to the amount of nitrifying bacteria present. There are several operational factors that determine the population size and activity of nitrifying bacteria and consequently, the ability of the activated sludge process to nitrify (Table 4.3). The operational factors are also the limiting factors.

Table 4.3 Favorable Operational Factors for Nitrification in the Mixed Liquor

Operational Factor	Operational Condition
Alkalinity	7.14 mg/L × $TKN_{influent}$ mg/l + 50 mg/L
Dissolved oxygen	2 to 3 mg/L
F/M	< 0.08
Inhibition/toxicity	Absent
MLVSS	> 2000 mg/L
pH	7.2 to 7.4
Soluble cBOD	< 15 mg/L
Temperature	> 16°C

The growth rate of nitrifying bacteria increases during warm wastewater temperature and decreases during cold wastewater temperatures (Table 4.4). From 5 to 20°C NOB grow faster than AOB, while AOB grow faster than NOB between 20 to 40°C. Therefore, at temperatures > 20°C the oxidation of nitrite to nitrate is the rate-limiting reaction, and at temperatures < 20°C the oxidation of ammonium to nitrite is the rate-limiting reaction. Minimum detention time for nitrification is approximately four hours at 20 to 25°C.

Table 4.4 Guideline MCRT Needed for Nitrification in Temperate Regions of North America

Temperature °C	MCRT
25	10
20	15
15	20
10	30

Although nitrifying bacteria live longer than most heterotrophic bacteria, they have a longer generation time than most heterotrophic bacteria. Nitrifying bacteria have a long generation time (eight to ten hours) under optimal conditions in a laboratory and an even longer generation time (three to six days) in the activated sludge process. Therefore the number of nitrifying bacteria needed for successful nitrification must be increased during cold temperatures by decreasing wasting and increasing the mean cell residence time (MCRT) and mixed liquor volatile suspended solids (MLVSS). MLVSS represent the relative population size of bacteria. If the secondary clarifier cannot handle the increased solids loading, the use of fixed film or suspended media in the aeration tank would provide for the development of a biofilm having not only nitrifying bacteria but also cBOD-removing bacteria. The use of fixed film media in the activated sludge process is called integrated fixed film activated sludge (IFFAS).

During cold wastewater temperature nitrifying bacteria and heterotrophic bacteria both display decreased ability to degrade nBOD and cBOD, respectively. Therefore, an increase in MCRT results in an increase in not only the nitrifying bacterial population but also the heterotrophic bacterial population. A higher MCRT is needed for not only cold wastewater temperature nitrification but also consistent nitrification for difficult-to-degrade nitrogenous wastes. Higher MCRT values should be accompanied with a lower food-to-microorganism (F/M) ratio.

Chapter 5

Nitrification and Limiting Factors

Nitrifying bacteria are more sensitive to upset than other bacteria in the activated sludge process, thus nitrification requires careful and often daily monitoring and process control. Successful nitrification is reflective of successful operation of the treatment process.

In addition to depressed temperature there are operational conditions or limiting factors that influence the population size of nitrifying bacteria and consequently the rate of nitrification. Depressed temperatures and limiting factors can cause the occurrence of incomplete nitrification or loss of nitrification. Therefore, it is necessary for the operator to monitor limiting factors and make appropriate adjustments as soon as possible for those factors that are responsible for incomplete nitrification.

Because nitrifying bacteria do not obtain much energy from the oxidations of ammonium and nitrite, they very quickly become sluggish or inactive when a factor is not acceptable, and they usually take two to three days to return to acceptable activity once the factor has been corrected. Limiting factors include: 1) alkalinity and pH, 2) dissolved oxygen concentration, 3) hydraulic retention time, 4) inhibition and toxicity including substrate toxicity, 5) organic loading and cBOD-to-TKN ratio, 6) orthophosphate concentration, 7) population of nitrifying bacteria, and 8) slug discharge of soluble cBOD.

Although limiting factors cause a rapid change in the activity and number of nitrifying bacteria, an appropriate corrective measure does not result in a rapid recovery of the population. Due to the long generation time and small energy obtained from the oxidations of ammonium and nitrite, three days often are required for the population to recover. The time to recover during cold wastewater temperature is longer.

Temperature

The nitrification process is dependent on temperature. Nitrification can occur over a wide range of temperature values (8 to 45°C). Increasing wastewater temperature increases the growth rate of nitrifying bacteria and consequently, the rate of nitrification. Decreasing wastewater temperature decreases the growth rate of nitrifying bacteria and consequently, the rate of nitrification. Critical temperature values or range of temperature values are: 45°C, 30°C, 25 to 20°C, 15°C, 10°C, and 8°C (Figure 5.1).

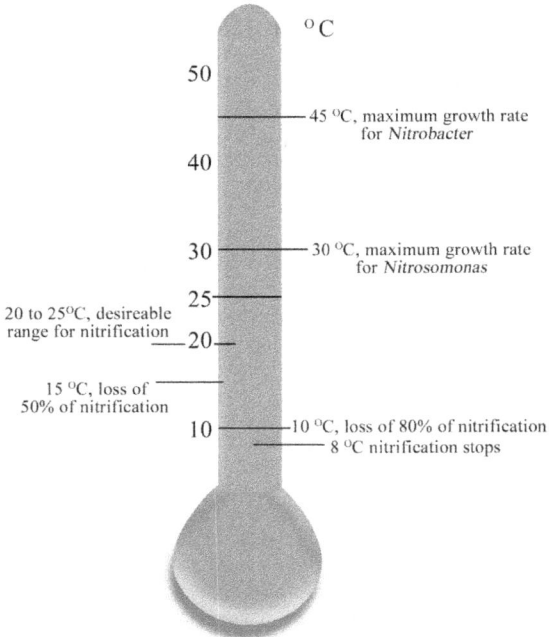

°C

50

— 45 °C, maximum growth rate for *Nitrobacter*

40

30 — 30 °C, maximum growth rate for *Nitrosomonas*

25

20 to 25°C, desireable range for nitrification — 20

15 °C, loss of 50% of nitrification

10 — 10 °C, loss of 80% of nitrification
— 8 °C nitrification stops

Figure 5.1 Temperature and its impact upon nitrification: Temperature has a profound impact upon the population size of the nitrifying bacteria and the activity of the nitrifying bacteria. Increasing temperature favors the growth of nitrifying bacteria and promotes the activity of the bacteria. Decreasing temperature hinders the growth of nitrifying bacteria and decreases the activity of the bacteria. There are five critical temperature values or ranges in values that affect nitrification. The temperature values of concern are: 1) at < 8°C nitrification stops; 2) at 10°C approximately 80% of the ability of the treatment process to nitrify is lost; 3) at 15°C approximately 50% of the ability of the treatment process to nitrify is lost; 4) nitrification is best from 20 to 25°C; 5) at 30°C the maximum growth rate for Nitrosomonas occurs; and 6) at 45°C the maximum growth rate for Nitrobacter occurs.

At 30°C some nitrifying enzymes are denatured or damaged. The optimum temperature for nitrification is approximately 25°C. Often little temperature effect occurs from 15 to 20°C. At 15°C approximately 50% of the ability of the activated sludge process to nitrify is lost. At 10°C nitrification is severely inhibited, and approximately 80% of the ability of the activated sludge process to nitrify is lost at 15°C. Nitrification stops at 5°C. Therefore, it is necessary to monitor and anticipate depressed temperature and implement timely and appropriate operational measures to maintain acceptable nitrification during cold-weather months. Operational measures to improve cold-weather nitrification should be implemented well before temperature drops to 15°C. Operational measures that can be used to overcome depressed temperature include many that are used to correct for limiting factors.

Limiting factors

Alkalinity

Alkalinity performs several roles in the activated sludge process. It buffers the pH of the mixed liquor, buffers cellular enzyme systems, and serves as the carbon substrate for nitrifying bacteria. If sufficient alkalinity is not present, the pH of the treatment process drops and nitrification stops. If there is no alkalinity, there is no carbon for the growth of nitrifying bacteria. If there is no growth, there is no need for energy. Therefore, oxidations of ammonium and nitrite stop.

Because nitrifying bacteria are obligate autotrophs, they can only use inorganic carbon, carbon dioxide (CO_2), as their carbon source. Bicarbonate (HCO_3^-) and carbonate (CO_3^{2-}) are the preferred carbon sources. Once absorbed by nitrifying bacteria, the bicarbonate and carbonate are hydrolyzed to produce carbon dioxide (Equations 5.1). There is a risk that in low-alkalinity wastewaters nitrification will lower the pH to an inhibitory level.

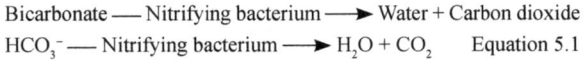

Bicarbonate ⎯⎯ Nitrifying bacterium ⎯⎯▶ Water + Carbon dioxide

HCO_3^- ⎯⎯ Nitrifying bacterium ⎯⎯▶ $H_2O + CO_2$ Equation 5.1

Alkalinity is lost in the sewer system by two means. First, industries may discharge wastewater that contains significant acidic wastes that neutralize alkalinity in the wastewater. Second, industries that discharge wastewater that contains significant alkalinity may reduce the quantity and strength of their wastewater. Dairy production is one industry that discharges wastewater with a relatively large quantity of alkalinity. Typically, alkalinity in untreated domestic wastewater is approximately 200 mg/L as calcium carbonate ($CaCO_3$)

If needed, alkalinity can be added to the activated sludge process with a variety of alkali compounds (Table 5.1). Some of these compounds, such as sodium hydroxide (NaOH), do not contain carbon. They contain the hydroxyl group (-OH). Once dissolved in wastewater they produce carbonate when the hydroxyl group bonds to dissolved carbon dioxide (Equation 5.2).

Table 5.1 Alkali Compounds Commonly Used for Alkalinity Addition

Compound	Common Name	Formula	Equivalence to $CaCO_3$
Calcium carbonate	Calcite	$CaCO_3$	1.00
Calcium bicarbonate	Calcium hydrogen carbonate	$Ca(HCO_3)_2$	0.62
Calcium hydroxide	Soda ash	$Ca(OH)_2$	1.35

Compound	Common Name	Formula	Equivalence to $CaCO_3$
Calcium oxide	Quicklime	CaO	1.80
Magnesium bicarbonate	Magnesium hydrogen carbonate	$Mg(HCO_3)_2$	0.68
Magnesium carbonate	Magnesite	$MgCO_3$	1.19
Magnesium hydroxide	Magnesia, Mag	$Mg(OH)_2$	1.13
Sodium bicarbonate	Baking soda	$NaHCO_3$	0.60
Sodium carbonate	Soda ash	Na_2CO_3	0.94
Sodium hydroxide	Caustic soda	NaOH	1.25

Sodium hydroxide + Carbon dioxide $\longrightarrow$ Sodium carbonate + Water

$$2NaOH + CO_2 \longrightarrow Na_2CO_3 + H_2O \qquad \text{Equation 5.2}$$

Nitrification consumes alkalinity and lowers pH. This occurs because: 1) nitrous acid (HNO_2) is produced in the first step of nitrification (Equations 5.3 and 5.4), and 2) some alkalinity is consumed as a carbon source for bacterial growth. Nitrous acid destroys alkalinity. In the first step of nitrification nitrite (NO_2^-) and hydrogen protons (H^+) are produced. The bonding of nitrite with a hydrogen proton produces nitrous acid. With the loss of alkalinity and buffering ability, pH drops in the aeration tank.

Ammonium + Oxygen ── AOB ──▶ Protons + Nitrite + Water

$NH_4^+ + 1.5\ O_2$ ── AOB ──▶ $2\ H^+ + NO_2^- + H_2O$ Equation 5.3

(first step in nitrification)

Proton + Nitrite ──▶ Nitrous acid

$H^+ + NO_2^-$ ──▶ HNO_2 Equation 5.4

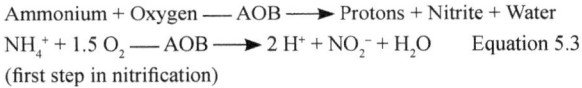

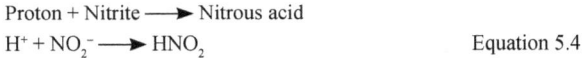

(production of nitrous acid from protons and nitrite from the first step of nitrification)

Alkalinity is reported as calcium carbonate ($CaCO_3$). Alkalinity may be measured in mg/L or lbs., and alkalinity needed may be calculated in mg/L or lbs. (Figure 5.2). The quantity of alkalinity needed should be based on influent TKN. Overall the theoretical alkalinity consumed during nitrification is 7.14 mg/L as calcium carbonate per mg/L of ammonium-nitrogen oxidized to nitrate-nitrogen or 7.14 lbs. as calcium carbonate per lb. of ammonium-nitrogen oxidized to nitrate-nitrogen.

> Given : Influent flow = 5MGD
>
> Influent TKN = 40 mg/L

Alkalinity consumed by nitrification:

In mg/L

| (40 mg/L | X | 7.14) | + 50 mg/L = | 336 |

(40 mg/L
TKN
concentration

X   7.14)
mg/L alkalinity
required per
mg/L of
ammonium – N
nitrified
to
nitrate

+ 50 mg/L =

336
mg/L
alkalinity
as
CaCO₃
required

In lbs. / day

(5 MGD x 40 mg/L x 7.14 x 8.34) + 50 lbs. = 11,959

Flow   TKN
concentration

lbs./
gallon

lbs.
alkalinity
as CaCO₃
per day

mg/L alkalinity
required per
mg/L of
ammonium – N
nitrified
to
nitrate

Figure 5.2 Calculating alkalinity needed for nitrification: The amount of alkalinity (carbon source) needed for nitrification can be calculated in mg/L or pounds (lbs.). Parameters needed for the calculation include influent TKN concentration, the constants for alkalinity (7.14 mg/L alkalinity required per mg/L of ammonium-nitrogen to be nitrified) and weight of wastewater (8.34 lbs./gallon), and the quantity of alkalinity (50 mg/L or 50 lbs.) to drive the reaction.

The quantity of ammonium-nitrogen actually consumed is slightly less than the theoretical, since some ammonium-nitrogen is absorbed from the wastewater and used as a nitrogen nutrient for growth. Absorbed ammonium-nitrogen that is used as a nutrient is not oxidized. It is assimilated into cellular material.

An extra 50 mg/L or lbs. of alkalinity is added to the calculated value of needed alkalinity in order to "drive" the first step of nitrification that consumes the alkalinity. If sufficient alkalinity is not present, inhibition of nitrification may occur. Nitrification is inhibited when alkalinity is < 40 mg/L.

If the quantity of alkalinity needed is less than the influent alkalinity, then alkalinity must be added to the activated sludge process. Alkalinity may also be considered adequate if the mixed liquor effluent filtrate demonstrates complete nitrification: ≥ 50 mg/L alkalinity, ≤ 1 mg/L NH_4^+-N, ≤ 1 mg/L NO_2^--N, and ≤ 1 mg/L NO_3^--N.

Although denitrification produces alkalinity, the net loss of alkalinity between nitrification (7.14 mg/L alkalinity consumed) and denitrification (3.6 mg/L alkalinity produced) may not be sufficient for nitrification. The chemical or chemicals selected for alkalinity addition should be determined by cost and logistics (safety, handling, storage, etc.). Some chemicals increase pH greatly, and some chemicals increase pH slightly.

Dissolved oxygen concentration

Proper dissolved oxygen concentration is critical for successful nitrification. Nitrifying bacteria are strict aerobes. They must have dissolved oxygen in order to be active and grow. Nitrifying bacteria are found on the surface as well as the interior of the floc particle. Therefore, increased diffusion of dissolved oxygen into the floc particle results in an increase in the rate of nitrification.

Nitrifying bacteria can survive without dissolved oxygen for approximately four hours. After four hours nitrifying bacteria begin to die. After 24 hours without dissolved oxygen, the population of nitrifying bacteria is considered to be dead. Because nitrifying bacteria are continuously recycled from the secondary clarifier to the aeration tank, excess sludge residence time in the clarifier should be avoided.

Species of AOB and NOB have significantly different oxygen sensitivity. In the activated sludge process nitrite oxidation to nitrate is more sensitive to low dissolved oxygen than AOB. As a result, nitrite may accumulate under low dissolved oxygen concentration. NOB have a lower affinity for dissolved oxygen than AOB, and the activity of NOB is limited more than the activity of AOB with decreasing dissolved oxygen.

The maximum growth rates for AOB and NOB also are affected by dissolved oxygen concentration. The dissolved oxygen concentration needed for maximum growth rates may be as low as 0.3 mg/L and as high as 4.0 mg/L or more. There are several operational factors that are responsible for the wide range of dissolved oxygen concentrations needed for maximum growth rates. These factors include: 1) the effects of dissolved oxygen diffusion in floc particles, 2) variations in measurement of dissolved oxygen, 3) MCRT, 4) dissolved oxygen saturation of the bulk solution, and 5) solids residence time (SRT). A higher MCRT results in a higher concentration of MLVSS (bacteria). An activated sludge process with a higher MCRT can achieve acceptable nitrification with a low dissolved oxygen concentration as compared to an activated sludge process with a lower MCRT.

Operational conditions for nitrification must also satisfy cBOD degradation. Given the large variation of wastewater flow, strength, and composition and operation factors previously addressed suggest that a specific dissolved oxygen concentration or specific range of concentrations does not exist.

As dissolved oxygen concentration increases, so does the rate of nitrification. Nitrifying bacteria often grow best between 2 to 3 mg/L of dissolved oxygen. Concentrations of dissolved oxygen > 3 mg/L also may improve nitrification, if heterotrophic bacteria can degrade cBOD more rapidly. This would give the nitrifying bacteria more time to nitrify. The growth of nitrifying bacteria occurs at an oxidation-reduction potential (ORP) approximately between +100 and +350 millivolts (mV) (Table 5.2).

Table 5.2 Guideline ORP Values and Cellular Activity

Cellular Activity	ORP, mV
Nitrification	+100 to +350
cBOD degradation with oxygen	+50 to +250
Biological phosphorus uptake	+25 to +250
Denitrification	+50 to −50
Sulfide production (sulfate reduction)	−50 to −250
Biological phosphorus release	−100 to −250
Acid formation (mixed acid production/fermentation)	−100 to −225
Methane production	−175 to −400

With increasing wastewater temperature (> 15°C), temperature may become the limiting factor. At temperatures > 15°C the warm wastewater has less affinity for dissolved oxygen. A reduction in oxygen transfer into the interior of the floc particle due to warm wastewater temperature decreases the rate of nitrification.

Nitrifying bacteria consume approximately 4.6 lbs. of oxygen for each pound of ammonium oxidized to nitrate. The first step in nitrification requires 3.43 pounds of oxygen for each pound of ammonium oxidized to nitrite. The second step in nitrification requires 1.14 lbs. of oxygen for each pound of nitrite oxidized to nitrate. For comparison, 1.6 to 1.8 lbs. of oxygen are consumed to oxidize 1 lb. of cBOD to cells (sludge), carbon dioxide, water, and other compounds such as ammonium, phosphate, and sulfate (SO_4^{2-}). The oxygen demand for complete nitrification may increase power requirements by 40% of that required for cBOD removal.

When cBOD is present, nitrifying bacteria must compete with the heterotrophic bacteria for dissolved oxygen. Since the heterotrophic bacteria have a higher growth rate (short generation time) and greater affinity for oxygen, the heterotrophic bacteria outcompete the nitrifying bacteria for dissolved oxygen. Therefore, nitrification occurs only when cBOD has been reduced to a low concentration.

Heterotrophic bacteria perform a key role in limiting oxygen transfer to nitrifying bacteria. Due to the higher growth rate of heterotrophic bacteria, the rate of nitrification is also controlled by the oxidation of cBOD. As long as there is a high cBOD loading, the heterotrophic bacteria will dominate. Therefore, nitrification systems must have sufficient detection time within the aeration tank for nitrifying bacteria to grow.

Hydraulic retention time

It is not necessary to have the maximum growth rate for nitrifying bacteria in order to obtain effective nitrification, if there is adequate hydraulic retention time (HRT) or contact time between substrate and nitrifying bacteria. Increasing HRT provides more time for nitrification after cBOD degradation has been satisfied. Increasing HRT also provides more retention time within the activated sludge process for nitrifying bacteria, especially at high cBOD loading.

Inhibition and toxicity including substrate toxicity

The terms "inhibition" and "toxicity" are used interchangeably. For the purpose of this book, inhibition is damage to the bacterial cell that results in sluggish activity or temporary loss of activity. Inhibition may be acute or chronic. If time and sufficient energy are available to the bacterium, the bacterium can recover from inhibition. Toxicity results in the death of a bacterial cell. Toxicity may be acute or chronic. In addition, AOB and NOB have different degrees of tolerance to specific inhibitory and toxic compounds and ions found in industrial wastewaters. The difference in tolerance can result in nitrite production and accumulation.

Generally, whatever is inhibitory or toxic to heterotrophic bacteria is also inhibitory or toxic to nitrifying bacteria. However, inhibition or toxicity occurs at a lower concentration of an inhibitory or toxic organic compound or inorganic compound or metal ions such as copper (Cu^{2+}) and hexavalent chromium (Cr^{6+}). This susceptibility of nitrifying bacteria to inhibition or toxicity at a lower concentration is due to the relatively small amount of energy that nitrifying bacteria have available to repair cellular damage as compared to the energy available for cellular repair in cBOD-removing bacteria.

Over 150 organic compounds and over 30 inorganic compounds and metal ions have been identified as being inhibitory or toxic to nitrifying bacteria. More commonly cited inhibitory or toxic compounds and ions are listed in Table 5.3.

Table 5.3 Examples of Inhibitory or Toxic Compounds or Ions to Nitrifying Bacteria

Group	Compound or Ion
Organic	Acetone
	Aniline
	Carbon disulfide
	Chloroform
	Ethanol
	Ethylenediamine
	Hexamethylene diamine
	Monoethanolamine
	Phenol

Group	Compound or Ion
Inorganic	Arsenic
	Cadmium
	Chromium
	Cobalt
	Copper
	Cyanide
	Fluoride
	Free ammonia
	Hydrazine
	Lead
	Mercury
	Nickel
	Nitrous acid
	Perchlorate
	Potassium chromate
	Silver
	Sodium azide
	Sodium cyanate
	Thiocyanate
	Zinc

Therefore, in order to prevent inhibition and toxicity, industrial dischargers of inhibitory and toxic compounds and ions should be identified and regulated. In addition to source control at industrial sites there are some operational measures that can be used to prevent inhibition or toxicity in the activated sludge process. These measures include:

• using complete mix rather than plug-flow mode of operation in order to dilute influent toxicant between tanks rather than concentrate the toxicant in one tank (Figure 5.3)

- decreasing the toxic mass-to-biomass ratio by increasing MLVSS
- using bio-augmentation cultures that can degrade inhibitory or toxic organic compounds or safely bio-accumulate metal ions
- controlling pH

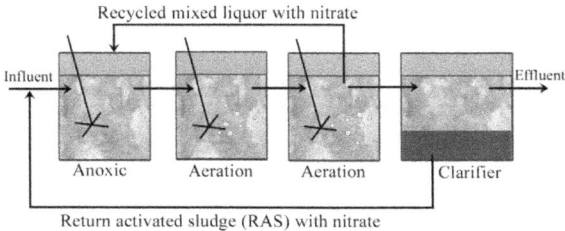

Figure 5.3 Plug-flow mode of operation: In plug-flow mode of operation all of the influent enters the first reactor. In this schematic the first reactor is an anoxic (denitrifying) reactor. In plug-flow mode of operation all the toxic waste in the wastewater also enters the first reactor. This mode of operation offers the opportunity for the wastewater to cause toxicity.

Inhibition and toxicity may occur in-plant. The accumulation of nitrite and nitrate can cause inhibition and toxicity. Nitrite accumulation can inhibit AOB. Nitrate accumulation > 20 mg/L can cause failure of the whole effluent toxicity test (WETT). At elevated pH (> 8.0) free ammonia is inhibitory to NOB at concentrations from 0.1 to 1.0 mg/L and is toxic at 3.0 mg/L to aquatic life. A long-term safe concentration of ammonia of 0.02 mg/L is most often used.

In addition to toxic compounds and ions there are two unique forms of toxicity that occur only with nitrifying bacteria. These are "recognizable," soluble cBOD toxicity (Table 5.4) and substrate toxicity. Recognizable, soluble cBOD consists of a relatively small number of simplistic, highly soluble compounds, mostly alcohols that are "recognized" by enzyme systems in nitrifying bacteria. Once they are recognized, the enzyme systems shut down. It is not until these compounds are reduced in quantity or degraded completely that enzymatic activity returns. Heterotrophic bacteria degrade recognizable, soluble cBOD compounds.

Table 5.4 Alcohols That Are Recognizable, Soluble cBOD Compounds

Alcohol	Formula
Methanol	CH_3OH
Ethanol	$CH3CH_2OH$
Propanol	$CH_3CH_2CH_2OH$
Butanol	$CH_3(CH_2)_2CH_2OH$

Substrate toxicity occurs when the concentration of the energy substrate ammonium in the aeration tank is > 500 mg/L (Figure 5.4). Deamination and ammonification of organic nitrogen compounds are responsible for a higher concentration of ammonium in the mixed liquor as compared to the concentration of ammonium in the mixed liquor influent.

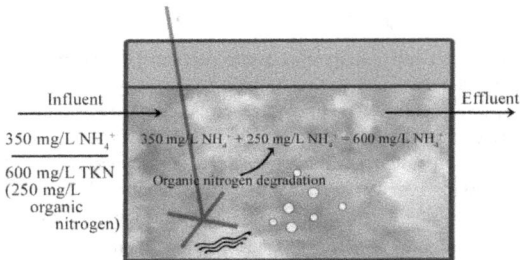

Figure 5.4 Concentration of ammonium in mixed liquor influent and mixed liquor: The concentration of ammonium in the mixed liquor can be higher than the concentration of the ammonium in the mixed liquor influent. This would occur due to the deamination and ammonification of organic nitrogen compounds in the mixed liquor tank. If the TKN of the mixed liquor is not measured, then the influent ammonium concentration may result in substrate toxicity in the mixed liquor.

At high concentrations of ammonium in the mixed liquor, one of the following may occur: 1) AOB may not be able to efficiently oxidize ammonium to nitrite, resulting in a high pH that causes free ammonia toxicity, or 2) AOB can oxidize ammonium to nitrite at low pH, but NOB cannot oxidize nitrite to nitrate, resulting in a low pH and causing free nitrous acid toxicity.

Substrate toxicity may be prevented by regulating discharges of nitrogenous compounds to ensure a safe level of ammonium in the mixed liquor tank. In addition, the maintenance of a near-neutral pH in the mixed liquor is helpful in preventing substrate toxicity.

Organic loading and cBOD-to-TKN ratio

The heterotrophic bacteria perform positive and negative roles related to population size and activity of nitrifying bacteria. Positive roles performed by heterotrophic bacteria include: 1) reduction or removal of inhibitory or toxic wastes including recognizable, soluble cBOD, and 2) development of floc particles that serve as a site for the attachment of nitrifying bacteria. Nitrifying bacteria are poor floc-forming organisms.

A negative role is the overgrowth of heterotrophic bacteria. Heterotrophic bacteria grow much faster than nitrifying bacteria at high cBOD loadings. At high cBOD loadings, heterotrophic bacteria outgrow nitrifying bacteria and cause the nitrification process to fail. The large population of heterotrophic bacteria consumes large quantities of dissolved oxygen and nutrients.

A major factor that affects the ratio of heterotrophic bacteria to nitrifying bacteria in the mixed liquor influent is the cBOD-to-TKN ratio (Table 5.5). Because many activated sludge processes are designed for the higher growth rate of heterotrophic bacteria, the MCRT or MLVSS must be increased to provide for a larger inventory of nitrifying bacteria in order to provide for successful nitrification. However, increasing the MCRT or MLVSS may result in the following operational conditions:

- undesired growth of filamentous organisms
- foam production

- endogenous respiration
- overloading of the secondary clarifier

Table 5.5 cBOD:TKN Ratio and Fraction of Nitrifying Bacteria in MLVSS

cBOD:TKN Ratio	Fraction of Nitrifying Bacteria	Nitrification Rate
> 3.0	Low	Low
2.0 to 3.0	Moderate	Moderate
< 2.0	High	High

By lowering the cBOD-to-TKN ratio entering the mixed liquor the heterotrophic bacterial population is lowered. This change permits the nitrifying bacterial population to better compete with the heterotrophic population for dissolved oxygen and nutrients.

Recycle streams, especially from solids handling and dewatering processes, potentially may be large sources of concentrated ammonium and organic nitrogen (Table 5.6). These streams, if discharged to the mixed liquor, increase the demand for oxygen and alkalinity in order to nitrify and carbon to denitrify. Therefore, the timing of solids handling and dewatering processes should be scheduled during low loading conditions. Periodic or daily monitoring of the mixed liquor influent TKN, ammonium, nitrite, and nitrate are needed in order to obtain a working hourly profile of influent nitrogenous compounds.

Table 5.6 Side Stream Processes (Recycle Streams) That Have Nitrogen and Phosphorus

Aerobic digester decant
Anaerobic digester decant
Belt filter press filtrate
Centrifuge centrate
Dissolved air filtration subnatant

Gravity belt thickening filtrate
Gravity thickening supernatant
Plate and frame filtrate
Reed bed filtrate/leachate

Orthophosphate concentration

Orthophosphate ($H_2PO_4^-$) is the only form of phosphorus that can be used as the phosphorus nutrient by bacteria. An adequate quantity of orthophosphate must be present in the mixed liquor to satisfy the phosphorus needs for heterotrophic bacteria and nitrifying bacteria. To ensure that an adequate quantity of orthophosphate is available, a grab sample of mixed liquor effluent during peak loading conditions should be obtained and filtered. The filtrate should be tested for orthophosphate-phosphorus. If the concentration is ≥ 0.5 mg/L, then an adequate quantity of phosphorus is available provided there is no toxicity in the mixed liquor. If phosphorus is needed there are many phosphorus-containing compounds that can be used to add phosphorus to the activated sludge process; phosphoric acid (H_3PO_4) is most commonly used.

Before adding phosphorus to the activated sludge process the following operational measures should be performed:

- if orthophosphate is being precipitated in the primary clarifier or mixed liquor to satisfy a total phosphorus (TP) discharge limit, consider changing the precipitation site to the mixed liquor effluent or secondary clarifier influent.
- if orthophosphate is being precipitated at the beginning of the react phase of a sequencing batch reactor (SBR) to satisfy a total phosphorus discharge limit, consider changing the precipitation time to the end of the react phase.
- if the concentration of the orthophosphate in the mixed liquor effluent or end of react phase is targeted for ≥ 0.5 mg/L to prevent a phosphorus deficiency, then excess phosphorus must be precipitated in order to satisfy a total phosphorus discharge limit.

pH

An acceptable pH value for nitrification does not indicate an acceptable level of alkalinity, and an acceptable level of alkalinity does not indicate an acceptable pH. For example, the influent to a wastewater treatment plant may have pH 7.0 and an alkalinity value of 200 mg/L. The pH of an anaerobic digester may also have pH 7.0 and an alkalinity of 2000 mg/L. Therefore pH indicates where the mixed liquor is with regard to an acceptable pH. Daily or periodic testing for alkalinity reveals the trend of alkalinity over time in the mixed liquor. A decreasing trend indicates a potential problematic condition and immediate process-corrective action. If the wastewater has a relatively low alkalinity concentration, a change in pH can be rapid and unacceptable. This is detrimental, because pH affects enzyme structure and enzymatic activity.

Nitrifying bacteria obtain little energy from the oxidations of ammonium and nitrite and have difficulty in adjusting to changes in operational conditions. Nitrifying bacteria need a steady-state pH with a $\pm$ 0.3 pH change on a daily basis. Typically, an operating range of pH from 7.2 to 7.4 is optimal for nitrification. However, many activated sludge processes operate at a slightly lower range of pH values, from 6.8 to 7.2.

Nitrifying bacteria are capable of growing over a wide range of pH values, 6.5 to 8.0. However, below pH 7.2 the rate of nitrification decreases rapidly and approaches zero at pH 6. At pH values < 6.5 or > 8.0, damage to enzymes occurs resulting in inhibition.

Population of nitrifying bacteria

Nitrifying bacteria can be added to an activated sludge processors by three means. First, the plant receives nitrifying bacteria in large numbers on a continuous basis from I/I. Natural seeding (I/I) may be all that is required for the addition of nitrifying bacteria. However, because nitrifying bacteria are strict aerobes and are found only in the top two inches of soil, natural seeding may be ineffective where the soil freezes below two inches. Second, nitrifying bacteria may be added to the activated sludge process by introducing mixed liquor from another activated sludge process that is nitrifying and receives similar influent. Third, they may also be added through bio-augmentation. Here commercially grown cultures of nitrifying bacteria are added to the mixed liquor. Because nitrifying bacteria grow slowly, are difficult to culture, and do not form spores, the cost of the cultures may be several hundred dollars per gallon. Spores protect bacteria from harsh environmental conditions including desiccation. Because the cost for commercially prepared cultures of nitrifying bacteria is expensive, the following measures may be used to reduce the cost and promote acceptable bacterial activity:

- test the mixed liquor effluent filtrate for the presence of nitrite or nitrate.
- if nitrite or nitrate is present, then a nitrifying bacterial population also is present. To promote the growth of the population add saprophytic (heterotrophic) bacteria to the mixed liquor to remove cBOD more rapidly to provide more time in the aeration tank for the existing population of indigenous nitrifying bacteria to nitrify.
- if nitrite or nitrate is not present, then an active population of nitrifying bacteria also is not present. To augment the mixed liquor with nitrifying bacteria add saprophytic bacteria. The addition of saprophytic bacteria again will also remove cBOD more rapidly, thus providing more time for the augmented population of nitrifying bacteria to grow and oxidize ammonium and nitrite.

Slug discharge of soluble cBOD

A slug discharge of soluble cBOD may be defined as a cBOD loading two to three times greater than normal loading over three to four consecutive hours. A slug discharge may cause the following operational conditions that adversely impact nitrification: 1) reduction in hydraulic retention time (HRT), 2) low dissolved oxygen level, 3) nutrient deficiency, and 4) presence of recognizable, soluble cBOD. To prevent a slug discharge of soluble cBOD, industries with the potential to cause such a discharge should be identified and regulated.

There may be more than one limiting factor responsible for the loss of nitrification, and a limiting factor may occur during cold wastewater temperatures. Therefore a checklist of temperature and limiting factors may be used to ensure that temperature and limiting factors are monitored on a regular basis or more often during incomplete nitrification (Table 5.7).

Table 5.7 Checklist for Identifying Temperature and Limiting Factors Responsible for the Occurrence of Incomplete Nitrification

if Monitored	Temperature or Limiting Factor
	Acceptable wastewater temperature
	Adequate alkalinity
	Adequate dissolved oxygen concentration
	Acceptable ORP value
	Acceptable HRT in mixed liquor tank
	Absence of inhibition and toxicity
	Acceptable mode of operation
	Acceptable concentration of ammonium in the mixed liquor tank
	Acceptable concentration of orthophosphate in mixed liquor effluent filtrate
	Acceptable and steady-state pH
	Abundant and active population of nitrifying bacteria
	Acceptable cBOD-to-TKN ratio

Commonly occurring operational conditions responsible for the loss of nitrification or lack of complete nitrification as well as a violation of the total nitrogen discharge limit include the following: 1) low dissolved oxygen level, 2) lack of alkalinity, 3) excess cBOD containing organic nitrogen (and organic phosphorus), 4) excess effluent TSS that contains organic nitrogen (and organic phosphorus), and 5) lack of orthophosphate.

Low dissolved oxygen level

The following are nonmechanical operational conditions that contribute to low dissolved oxygen levels in the aeration tank:

- excess or slug discharge of soluble cBOD
- presence of oxygen scavengers including sulfide (HS^-) and sulfite (SO_3^{2-})
- occurrence of endogenous respiration
- stratification of dissolved oxygen

Lack of alkalinity

The following are operational conditions that contribute to a lack of adequate alkalinity or change in alkalinity in the aeration tank:

- excess quantity of soluble cBOD is degraded in the aeration tank resulting in the release of much carbon dioxide and the formation of carbonic acid.
- excess quantity of nBOD is degraded in the aeration tank. The degradation results in the destruction of much alkalinity.
- sulfide (HS^-) oxidation has occurred in the aeration tank. Sulfide oxidation results in the loss of alkalinity.
- excess acidic compounds used at the treatment plant enter the aeration tank.
- acidic compounds used at industries are discharged to the sewer system.
- industrial discharges containing alkalinity are terminated, or the industrial discharges contain less alkalinity than normally discharged.

Excess cBOD and TSS containing organic nitrogen

The following are operational conditions that contribute to excess cBOD or TSS containing organic nitrogen in the final effluent:

- accumulation of floatable solids
- decrease in HRT
- significant change in F/M
- increase in organic loading
- undesired pH
- depressed temperature
- low dissolved oxygen level
- nutrient deficiency
- over-aeration and shearing of solids
- inhibition or toxicity
- poorly settling solids
- foam and scum accumulation
- Zoogloeal growth (viscous floc)
- lack of ciliated protozoa
- septicity
- surfactants
- undesired filamentous organism growth

Lack of orthophosphate in the mixed liquor

The following are operational conditions that contribute to lack of phosphorus in the mixed liquor:

- nutrient deficiency caused by an industrial discharge
- excess precipitation of orthophosphate upstream of the aeration tank

Chapter 6

Promoting Nitrification

In addition to the values or range of values that are commonly recommended to achieve and maintain acceptable nitrification (Table
6.1), there are several operational measures that can be used to promote nitrification (Table 6.2). These measures include: 1) increasing
dissolved oxygen concentration, 2) increasing hydraulic retention
time (HRT), 3) increasing the population size of nitrifying bacteria,
4) maintaining adequate influent alkalinity, 5) maintaining steady-
state pH in the aeration tank, 6) monitoring nitrogenous species, 7)
overcoming cold wastewater temperature, 8) preventing orthophosphate deficiency, 9) reducing cBOD loading to the aeration tank, 10)
reducing nitrifying bacteria washout, and 11) regulating discharge
of recycle streams. Techniques for achieving each measures are
checklisted in the following pages.

Table 6.1 Values or Range of Values for Parameters for Promoting Nitrification

Operational Parameter	Value/Range of Values
Alkalinity	7.14 mg/L × $TKN_{influent}$ mg/L + 50 mg/L
Dissolved oxygen	2 to 3 mg/L
F/M	< 0.08
MLVSS	> 2000 mg/L
ORP	+ 150 to + 350 mV
pH	7.2 to 7.4
Soluble cBOD	< 15 mg/L Mixed Liquor$_{effluent}$
Temperature	20 to 25°C

Table 6.2 Operational Measures to Promote Nitrification

Increasing dissolved oxygen concentration
Increasing hydraulic retention time (HRT)
Increasing the population size of nitrifying bacteria
Maintaining adequate influent alkalinity
Maintaining steady-state pH in the aeration tank
Monitoring nitrogenous species
Overcoming cold wastewater temperature
Preventing orthophosphate deficiency
Reducing cBOD loading to the aeration tank
Reducing nitrifying bacteria washout
Regulating discharge of recycle streams

Increasing dissolved oxygen concentration

- equalize flow of industrial discharges, especially those that contain relatively high quantities of soluble cBOD.

- identify sources of oxygen scavengers including sulfide (HS^-) and sulfite (SO_3^{2-}) and oxidize the scavengers before they enter the aeration tank or regulate the discharge of the scavengers. Sulfide may be oxidized using chlorine.

- increase the rate of aeration.

- perform dissolved oxygen profile to identify and correct stratification of dissolved oxygen and short-circuiting (Figure 6.1).

- prevent slug discharges of cBOD and nBOD.

- remove more cBOD in the primary clarifiers. See "reducing cBOD loading to the aeration tanks."

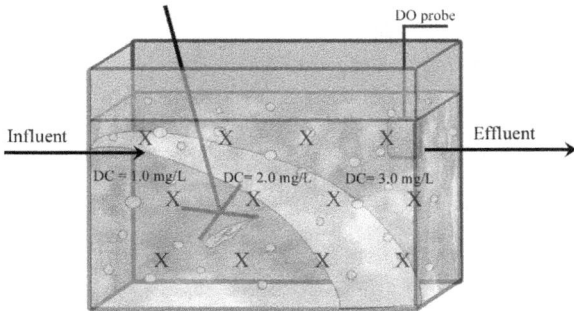

Figure 6.1 Dissolved oxygen profile and stratification: If short-circuiting occurs in the aeration tank, dissolved oxygen stratification may occur. Stratification results in zones of low and high dissolved oxygen concentrations. In zones of low dissolved oxygen, nitrification may be hindered or may not occur. This results in unacceptable nitrification or incomplete nitrification. To detect dissolved oxygen stratification, a dissolved oxygen profile of the aeration tank should be made by monitoring dissolved oxygen concentration in a grid-like fashion across and into the mixed liquor. Corrective measures for short-circuiting include draining and cleaning the tank and inserting baffles to correct for undesired flow patterns.

Increasing hydraulic retention time (HRT)

- place additional aeration tanks in-line.
- place additional primary clarifiers in-line.
- place additional secondary clarifiers in-line.
- reduce inflow and infiltration (I/I).
- reduce the rate of return activated sludge (RAS).
- thicken secondary clarifier solids with an appropriate coagulant or polymer and reduce the rate of return activated sludge (RAS).

Increasing the population size of nitrifying bacteria

- add commercially prepared cultures of saprophytic (heterotrophic) bacteria and nitrifying bacteria.
- avoid erratic sludge wasting rates, thereby preventing excess wasting of nitrifying bacteria and "pockets" of young growth in the mixed liquor that do not support the growth of slowly reproducing nitrifying bacteria.
- capture secondary clarifier suspended solids (bacteria) with an appropriate coagulant or polymer.
- insert fixed film or suspended media in aeration tanks to provide additional surface area for the attachment and growth of nitrifying bacteria.
- lower the wasting rate, if the following in the secondary clarifier do not occur: solids overloading, foam production, and undesired growth of filamentous organisms.

Maintaining adequate influent alkalinity

- identify industrial wastewaters with a low pH that are discharged to the sewer system. Equalize and neutralize the discharges to the sewer system to prevent destruction of alkalinity.
- identify industrial wastewaters with a relatively high alkalinity such as dairy wastewater. Monitor the discharges to determine when the quantity of alkalinity in the discharge decreases in order to quickly compensate for the reduction in alkalinity.

Maintaining steady-state pH in the aeration tank

- monitor and regulate chemical additions to the treatment process.
- monitor and regulate and if necessary neutralize industrial discharges.
- prevent + 0.3 change in pH on a daily basis.
- promote an acceptable pH range (7.2 to 7.4).

Monitoring nitrogenous species

- monitor the mixed liquor effluent filtrate from an in-line aeration tank for the following nitrogenous species: ammonium (NH_4^+), nitrite (NO_2^-), and nitrate (NO_3^-) to determine if complete or incomplete nitrification is occurring. Monitoring should be performed on a routine basis with more frequent monitoring occurring during depressed temperatures. The earlier that incomplete nitrification is detected the more quickly the cause for incomplete nitrification can be determined and corrected. Testing should be performed on the same aeration tank at approximately the same time to monitor daily changes in nitrification.
- monitor influent TKN to determine the quantity of alkalinity and dissolved oxygen needed for nitrification.

Overcoming cold wastewater temperature

- initiate cold wastewater temperature operational measures before temperature decreases to 16°C.
- increase dissolved oxygen concentration to at least 3.0 mg/L. See "increasing dissolved oxygen concentration."
- insert fixed film or suspended media to provide additional surface area for the attachment of heterotrophic bacteria and nitrifying bacteria.
- increased hydraulic retention time. See "increasing hydraulic retention time (HRT)."
- reduce mixed liquor influent cBOD. See "reducing cBOD loading to the aeration tanks."

Preventing orthophosphate deficiency

- to ensure an adequate quantity of the phosphorus nutrient orthophosphate for nitrifying bacteria, precipitation of orthophosphate to satisfy a total phosphorus discharge limit should be performed with the mixed liquor effluent rather than the primary clarifier influent or mixed liquor influent.

- if needed, phosphoric acid (HPO_4) may be added to the mixed liquor influent to satisfy a phosphorus deficiency.

Reducing cBOD loading to the aeration tank

- place additional primary clarifiers in-line.

- reduce the quantity of fats, oils, and grease (FOG) discharged to the sewer system.

- remove particulate and colloidal cBOD in the primary clarifiers by adding an appropriate coagulant or polymer to the primary clarifier influent.

- remove primary clarifier scum more frequently to prevent "breakthrough" of scum.

Reducing nitrifying bacteria washout

- add ammonium during reduced nBOD loading to aeration tanks to maintain an adequate population of nitrifying bacteria.

- decrease the cBOD-to-TKN ratio. Heterotrophic bacteria grow faster than nitrifying bacteria at high cBOD loading conditions.

- insert fixed film or suspended media in aeration tanks to provide additional surface area for the attachment and growth of nitrifying bacteria.

- reduce inflow and infiltration (I/I).

- rotate aeration tanks in-line and off-line (Figure 6.2).

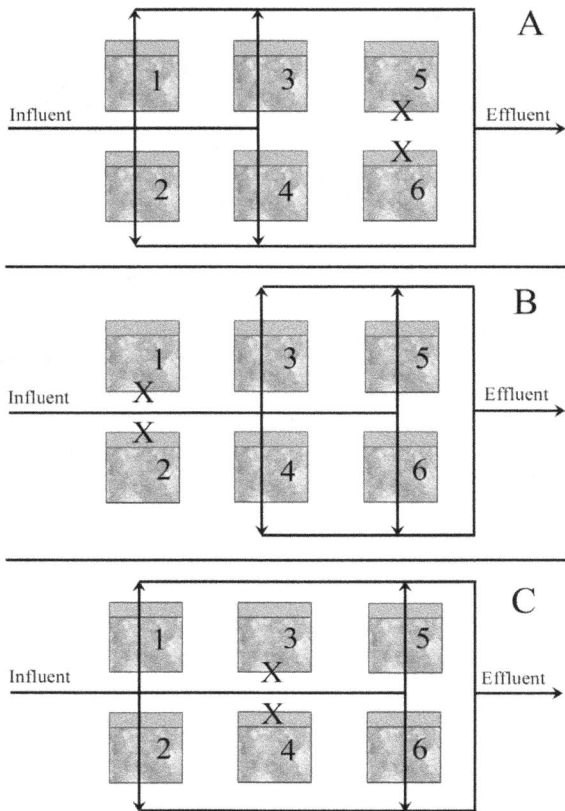

Figure 6.2 Rotating aeration tanks in-line and off-line: When hydraulic loading, carbonaceous loading, and nitrogenous loading permit, aeration tanks may be taken off-line on a rotating basis. For example, in a six-tank activated sludge process tanks 5 and 6 may be taken off-line (A) for a short period of time and then placed in-line while tanks 3 and 4 are taken off-line (B). Tanks

3 and 4 may be placed back in-line while tanks 1 and 2 are taken off-line (C). The rotation of aeration tanks in-line and off-line contributes to the following: 1) the washout of nitrifying bacteria in the off-line tanks, 2) the die-off of nitrifying bacteria in the in-line tanks when nitrogenous loading decreases to the activated sludge process, and 3) increase in F/M in the in-line tanks to overcome the undesired growth of low F/M growing filamentous organisms. As heterotrophic bacteria in the off-line tanks die due to the lack of carbonaceous loading, the dead cells undergo autolysis that releases organic nitrogen to the bulk solution, thus providing a continuous nitrogenous loading in the off-line tanks.

Regulating discharge of recycle streams

- monitor recycle streams for: 1) cBOD, 2) nBOD, 3) ammonium (NH_4^+), 4) nitrite (NO_2^-), 5) nitrate (NO_3^-), 6) organic nitrogen, 7) pH, and 8) alkalinity to determine the impact or loading of each parameter upon the treatment process to determine when the recycle stream should occur.

- monitor the following recycle streams: 1) centrate, 2) digester decant, 3) filtrate, 4) leachate from reed beds, 5) leachate from sludge-drying beds, 6) thickener overflow, and 7) septage or septage holding tanks to identify the strength of the nitrogenous loading from each stream.

PART TWO: DENITRIFICATION

Chapter 7

Denitrification: The Basics

Denitrification is the anaerobic respiration of soluble cBOD resulting in the reduction of nitrate (NO_3^-). Heterotrophic bacteria perform the reduction of nitrate in order to obtain carbon and energy for cellular growth from cBOD, resulting in the release of molecular nitrogen (N_2) and nitrous oxide (N_2) to the atmosphere. Since nitrogen leaves the bacterial cell it is not incorporated or assimilated into cellular material. This form of respiration is known as dissimilatory denitrification and is used at wastewater treatment facilities to satisfy a total nitrogen discharge requirement. Denitrification can include the use of nitrite (NO_2^-).

In the absence of ammonium (NH_4^+), the primary nitrogen nutrient for cellular growth, heterotrophic bacteria absorbed nitrate as the nitrogen nutrient. When this occurs nitrate is reduced in the cell, and nitrogen is incorporated into new cellular material. The use of nitrate as a nitrogen nutrient is known as assimilatory nitrate reduction. Nitrogen does not leave the cell.

Denitrification (where oxygen is absent) is the use of nitrate (NO_3^-) typically or nitrite (NO_2^-) atypically by facultative anaerobic or denitrifying bacteria to degrade soluble cBOD. Biological denitrification uses the ability of specific bacteria to use nitrate or nitrite for respiration under an anoxic condition (absence of oxygen).

Activated sludge processes denitrify for three reasons. The process is required to satisfy a total nitrogen discharge (TN). Therefore, the process must nitrify and then denitrify. The process is not required to satisfy a total nitrogen discharge limit, but the process is operated to nitrify and then denitrify in an anoxic reactor, selector, or zone in order to control the undesired filamentous organism growth (Table 7.1). The process is not required to satisfy a total nitrogen discharge limit and denitrification is not desired, but operational conditions allow the process to "slip" into denitrification. When an activated sludge process slips into denitrification, it usually occurs in the secondary clarifier. Here, large clumps of dark solids and bubbles rise to the surface. This condition is known as "clumping" or "dark sludge rising."

Table 7.1 Filamentous Organisms That Can Be Controlled in an Anoxic Condition

Haliscomenobacter hydrossis
Nocardioforms (*Nocardia*)
Nostocoida limicola
Sphaerotilus natans
Thiothrix spp.
Type 021N
Type 1701

Although there are several reactions in the reduction of nitrate to molecular nitrogen (N_2), the reduction of nitrate is typically present in two steps (Equations 7.1 and 7.2). The rates for the reactions are influenced by temperature.

Nitrate —— Denitrifying bacteria ——▶ Nitrite

NO_3^- —— Denitrifying bacteria —— NO_2^- Equation 7.1

Nitrite —— Denitrifying bacteria ——▶ Molecular nitrogen

NO_2^- —— Denitrifying bacteria ——▶ N_2 Equation 7.2

To satisfy the total nitrogen discharge limit an activated sludge process must nitrify and then denitrify. This requires the selection of an optimum range of operational conditions, perhaps seasonally adjusted for food-to-microorganism ratio (F/M) and MCRT in order to nitrify and denitrify successfully.

Nitrification does not remove nitrogen from the waste stream. It simply converts ammonium (NH_4^+) to nitrate (NO_3^-). Denitrification removes the nitrogen from the nitrate and releases the nitrogen to the atmosphere as mostly insoluble molecular nitrogen (N_2) and some nitrous oxide (N_2O) (Figure 7.1). Denitrification removes nitrogen from the wastewater. Except for nitrogen-fixing bacteria that fix molecular nitrogen to ammonium, the nitrogenous gases are not readily available for microbial growth.

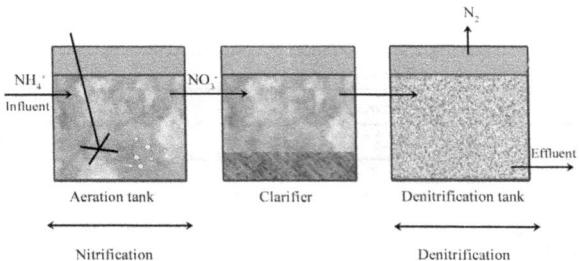

Figure 7.1 Removal of nitrogen from wastewater: Nitrogen is removed from the wastewater through its stripping to the atmosphere at high pH, assimilation in bacterial cells (sludge), and nitrification and denitrification. Nitrification occurs in the aeration tank where ammonium is oxidized to nitrate. Denitrification occurs in a denitrification filter or tank where nitrate is reduced to insoluble nitrogenous gases (molecular nitrogen and nitrous oxide). Nitrification does not remove nitrogen. It simply oxidizes reduced nitrogen (ammoniu~ oxidized nitrogen (nitrate). Nitrate-nitrogen remains in the wastewate~ trification reduces oxidized nitrate-nitrogen to nitrogenous gases. Ga~ tion results in the release of nitrogen to the atmosphere from the w~

In addition to molecular nitrogen and nitrous oxi~ dioxide is released to the atmosphere during denitrifica~ of nitrate to degrade cBOD does not yield as much ~ as the use of oxygen to degrade the same cBOD. V~ produced using nitrate for respiration there is le~ sludge yield). As compared to cBOD degradation~ dissolved oxygen (Equation 7.3), the carbon t'~

cellular growth during anoxic respiration (denitrification) goes into carbon dioxide (Equation 7.4). So much carbon dioxide is produced that it cannot all dissolve in the wastewater. The carbon dioxide that does not dissolve in the wastewater escapes to the atmosphere. Therefore, three gases are produced and released to the atmosphere during denitrification: molecular nitrogen, nitrous oxide, and carbon dioxide. There are four steps in the reduction of nitrate to molecular nitrogen (Equation 7.5).

1 lb. cBOD + DO $\longrightarrow$ **0.6 lbs.** Cells (sludge) + Water + Carbon dioxide

1 lb. cBOD + O_2 $\longrightarrow$ **0.6 lbs.** Cells (sludge) + H_2O + CO_2
$$\text{Equation 7.3}$$

1 lb. cBOD + Nitrate $\longrightarrow$ **0.4 lbs.** Cells (sludge) + Water + Carbon dioxide

1 lb. cBOD + NO_3^- $\longrightarrow$ **0.4 lbs.** Cells (sludge) + H_2O + $CO_2 \Uparrow$
$$\text{Equation 7.4}$$

Nitrate $\longrightarrow$ Nitrite $\longrightarrow$ Nitric oxide $\longrightarrow$ Nitrous oxide $\longrightarrow$ Molecular nitrogen

$NO_3^- \longrightarrow NO_2^- \longrightarrow N_2O \longrightarrow NO \longrightarrow N_2$ Equation 7.5

Denitrification involves the bacterial absorption of soluble cBOD and its degradation in an anoxic condition. Electrons released from the chemical bonds of the degraded cBOD are removed from the bacterial cell by bonding to nitrate (Figure 7.2). The removal of electrons results in the production of molecular nitrogen.

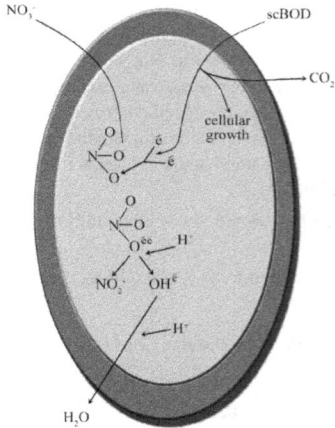

Figure 7.2 Removal of electrons by nitrate: In the absence of dissolved oxygen, denitrifying bacteria use nitrate for cellular respiration. Soluble cBOD is absorbed by the bacterium as a carbon and energy substrate. When cBOD is degraded inside the cell, carbon is released. Some of the carbon is used for cellular growth, while the remaining carbon leaves the cell as carbon dioxide. When the chemical bonds in the substrate are broken, electrons are released. The electrons give up some of their energy to the cell and ultimately the electrons must be transported out of the cell. To do this, nitrate is used as the electron transport molecule. The electrons attach to the oxygen atoms in nitrate, resulting in an increase in the negative charge of the oxygen atom. Positively charged hydrogen atoms or protons released from the degraded substrate are then bonded to the oxygen atom resulting in the release of the hydroxyl group (OH⁻), which combines with an additional proton to form water that is released from the cell. Oxygen is not produced during denitrification.

In order for denitrification to occur there must be an absence of dissolved oxygen or, if dissolved oxygen is present, an oxygen gradient must exist (Figure 7.3). Soluble, easily assimilated cBOD also must be present. The more simplistic in structure and soluble the cBOD is, the more easily it is degraded and the shorter the time period needed to remove any residual dissolved oxygen and reduce the quantity of nitrate. Denitrification usually requires one to four hours HRT in an anoxic zone or tank.

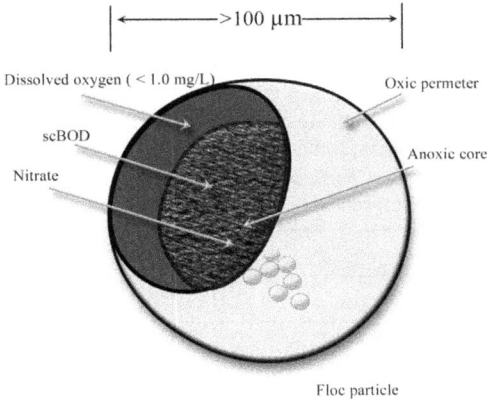

Floc particle

Figure 7.3 Oxygen gradient: Denitrification can occur in the presence of measureable dissolved oxygen. This can occur if the bulk solution contains nitrate, soluble cBOD, and < 1.0 mg/L dissolved oxygen, and the floc particle is > 100 µm in diameter. Bacteria on the perimeter of the floc particle use dissolved oxygen to degrade the cBOD. Because the relatively small quantity of dissolved oxygen is depleted as it moves toward the core of the floc particle and nitrate is not, nitrate is used in the core of the floc particle to degrade cBOD. In activated sludge processes that nitrify, a relatively high MCRT is needed to establish an abundant population of slow-growing nitrifying bacteria. The high MCRT promotes the growth of filamentous organisms. The organisms in turn provide strength to the floc particle, allowing the particle to overcome turbulence and grow in size.

There are four requirements that must be satisfied in order to denitrify. These requirements are:

- source of soluble cBOD (carbon source)
- absence of dissolved oxygen or presence of an oxygen gradient
- presence of nitrate (or nitrite)
- presence of an adequate and active population of denitrifying bacteria

Like nitrification, the critical requirement for denitrification is carbon. Alkalinity provides the carbon for the growth of nitrifying bacteria, while cBOD provides the carbon for the growth of cBOD-removing bacteria. Without carbon there can be no bacterial growth and consequently, there can be no nitrification or denitrification. Although there are several external organic compounds (not in the wastewater) that can be used as a carbon source for denitrification (Table 7.2), acetic acid and methanol have been studied the most. Methanol (CH_3OH) is commonly used.

Table 7.2 Organic Compounds Commonly Used as a Carbon Source for Denitrification

Compound	Formula
Acetic acid	CH_3COOH
Sodium acetate	CH_3COONa
Ethanol	CH_2H_5OH
Glucose	$C_6H_{12}O_6$
Methanol	CH_3OH

There are three sources of carbon that can be used for denitrification. The carbon sources include: 1) external, 2) internal, and 3) cellular. An external carbon source is the addition of chemicals such as acetic acid and methanol. The internal carbon source is the soluble cBOD in the influent wastewater, and the cellular source is the stored carbon in cellular starch granules and polysaccharides that coat the cell. Compared to internal and cellular carbon sources, external carbon sources have high removal rates for nitrate, and the denitrification rate is easily controlled through adjustment in chemical feed. Although acetic acid is used for denitrification, acetic acid added to the system neutralizes some alkalinity produced through denitrification. Atypical compounds or materials that have been used as carbon sources for denitrification include glycerol, lactic acid, molasses, peptone, and sawdust. Commercial products also are available for use as an external carbon source.

Denitrification may occur in an anoxic selector, an anoxic zone in an oxidation ditch, denitrification filter, dentrification tank, or in an aeration tank or aerobic digester with the cycling of air on (nitrification) and air off (denitrification) (Figure 7.4). Nitrification occurs within a range of redox values from + 100 to + 350 mV, while denitrification occurs within a range of redox values from + 50 to − 50 mV.

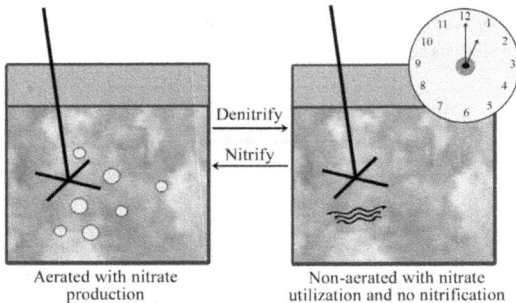

Figure 7.4 Denitrification in an aeration tank: Denitrification can occur in an aeration tank under the following conditions: 1) after soluble cBOD has been reduced to < 15 mg/L under aeration, nitrification occurs; 2) during nitrification, nitrates are produced; 3) after the production of nitrates, the tank is no longer aerated; and 4) after depletion of residual dissolved oxygen, the produced nitrate is then used to degrade soluble cBOD.

There are several modes of operation that are designed to nitrify and denitrify and include: 1) Bardenpho process, 2) Ludzack-Ettinger process, 3) Modified Ludzack-Ettinger process, and 4) Wahrmann process. Some modes of operation also remove phosphorus. Denitrification can be achieved in a sequencing batch reactor (SBR) with the use of a mixed fill phase (no aeration) or addition of a carbon source in the SBR at the end of settle phase and a short period of mixing after denitrification to strip entrapped gases in floating solids. Denitrification also can occur in a denitrification tank or filter downstream of the SBR. Nitrification is not restricted to a particular reactor; it occurs wherever there is an anoxic condition.

Significant indicators of denitrification in or across a treatment tank such as a denitrification tank or secondary clarifier include: 1) a decrease in nitrate or nitrite, 2) a decrease in oxidation-reduction potential (ORP), 3) an increase in alkalinity, and 4) an increase in pH (Figure 7.5). Additional indicators of denitrification may occur in a settleometer (Figure 7.6). Denitrification in the settleometer is the "first" of two "rise times" that occur in a settleometer, and indicators of the rise time include the presence of rising bubbles and dark sludge rising.

Influent			Effluent	
Alkalinity	Lower		Alkalinity	Higher
pH	Lower		pH	Higher
NO_3^-	Higher		NO_3^-	Lower
ORP	Higher		ORP	Lower

Figure 7.5 Indicators of denitrification across a secondary clarifier: If denitrification occurs in a secondary clarifier the following changes will occur with these specific parameters from influent to effluent of the aeration tank: 1) an increase in alkalinity due to the production of hydroxyl through denitrification, 2) an increase in pH as alkalinity is returned, 3) a decrease in nitrate as it is used for respiration, and 4) a decrease in oxidation-reduction potential (ORP) as oxidized nitrogen (nitrate) is reduced to nitrogenous gases (molecular nitrogen and nitrous oxide).

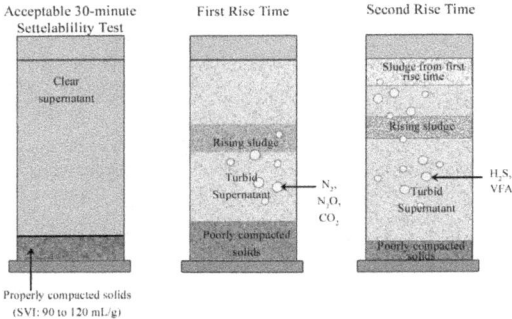

Figure 7.6 Rise times in a settleometer: If the 30-minute settleability test of nitrified mixed liquor is monitored over three hours, two rise times can be observed. The first rise time occurs approximately 1 to 1.5 hours after the start of the settleability test. The rise time occurs as residual dissolved oxygen in the mixed liquor is depleted and nitrate is used for bacterial respiration – denitrification. During denitrification three insoluble gases are produced and released. The gases consist of molecular nitrogen, nitrous oxide, and excess carbon dioxide. The gas bubbles as well as solids having entrapped gases rise to the surface of the mixed liquor. The second rise time occurs approximately two to three hours after the start of the settleability test. The rise time occurs as residual nitrate in the mixed liquor is depleted and sulfate reduction and fermentation occur. Sulfate reduction results in the production of insoluble hydrogen sulfide, while fermentation results in the production of volatile organic compounds, especially volatile fatty acids. The release of hydrogen sulfide and volatile organic compounds can be noted by the presence of gas bubbles rising to the surface of the mixed liquor.

Denitrification releases three insoluble gases: molecular nitrogen, nitrous oxide, and carbon dioxide. The gases or bubbles can be found rising to the surface of liquid medium or entrapped in solids rising to the surface.

The solids rising to the surface are dark in color. This is due to the accumulation of oils by old bacteria in the floc particles. In order to denitrify, the activated sludge process must first nitrify. Due to the long generation time to grow an adequate population of nitrifying bacteria, the activated sludge process usually is operated at a high MCRT.

While nitrification destroys alkalinity and decreases pH, denitrification returns alkalinity and increases pH (Equations 7.6). Alkalinity is returned to the treatment process by the production of the hydroxyl ion (OH^-). The hydroxyl ion reacts with dissolved carbon dioxide to produce bicarbonate alkalinity (Equation 7.7). Approximately half the alkalinity, 3.6 mg/L destroyed through nitrification, is returned to the treatment process through denitrification for each mg/L nitrate consumed for anoxic respiration (Figure 7.7). Recovery of alkalinity can help overcome low pH. The activity of denitrifying bacteria is affected by low pH. The activity of denitrifying bacteria and consequently the rate of denitrification are depressed at pH values < 6.0. Bacterial activity and rate of denitrification are also depressed at pH values > 8.0.

Carbon source + Nitrate —— Denitrifying bacteria ⟶

Cells (sludge) + Carbon dioxide + Molecular nitrogen + Hydroxyl

$cBOD + NO_3^-$ —— Denitrifying bacteria ⟶ Cells (sludge) + CO_2 + N_2 + OH^- Equation 7.6

Hydroxyl + Carbon dioxide ⟶ Carbonate + Water

$2\ OH^- + CO_2 \longrightarrow CO_3 + H_2O$ Equation 7.7

Given: Influent flow = 5 MGD
           Nitrate to be denitrified = 40 mg/L

Alkalinity produced by denitrification :

In lbs. / day

(5MGD   x   40 mg/L)   x   3.57   x   8.34   =

Flow   Nitrate to be   lbs. alkalinity   lbs. / gallon
            denitrified   per lb. nitrate
                               denitrified

5,955 lbs. alkalinity as $CaCO_3$
recovered per day

Figure 7.7 Calculating alkalinity produced through denitrification: The amount of alkalinity (carbon source) produced through denitrification can be calculated in mg/L or pounds (lbs.). To calculate pounds of alkalinity produced, the parameters needed for the calculation include the concentration of nitrate to be denitrified, the volume (MG or MGD) of nitrified wastewater to be denitrified, the constants for the return of alkalinity (3.57 lbs. alkalinity produced per lb. nitrate denitrified), and the weight of wastewater (8.34 lbs./ gallon).

Chapter 8

Denitrifying Bacteria

Denitrifying bacteria are heterotrophic, cBOD-degrading organisms. Denitrifying bacteria are also hardy facultative anaerobes and have significant differences in habitat, metabolism, and roles performed in the activated sludge process as compared to nitrifying bacteria (Table 8.1). As facultative anaerobes they are capable of using oxygen and nitrate (NO_3^-) to degrade cBOD (Table 8.2). However, they can only use oxygen *or* nitrate at any time. They use oxygen when oxygen and nitrate are both available. They use nitrate when it is available and oxygen is not available. The preference for oxygen is for the larger quantity of energy and growth obtained as compared to nitrate when degrading the same cBOD. Therefore, when oxygen and nitrate are available, oxygen "inhibits" the use of nitrate by denitrifying bacteria.

Table 8.1 Significant Differences in Habitat, Metabolism, and Roles Performed in the Activated Sludge Process

Parameter	Denitrifying Bacteria	Nitrifying Bacteria
Fate of nitrogen	Escapes to the atmosphere Used as a nitrogen nutrient	Remains in wastewater Energy substrate
Generation time	20 to 30 minutes	8 to 10 hours (laboratory) 3 to 6 days in treatment process
Carbon source for growth	Organic, soluble cBOD	Inorganic, carbon dioxide
Alkalinity	Produces alkalinity	Consumes alkalinity
Sludge yield	0.4 lb. sludge/lb. cBOD oxidize	0.12 lb. sludge/lb. ammonium oxidize
Respiration	Uses oxygen, nitrate, and nitrite	Uses dissolved oxygen
Optimum temperature for growth	20 to 25°C	20 to 25°C
Toxicity	Less sensitive	More sensitive
Habitat	Fecal waste, soil, water	Soil and water

Table 8.2 Comparison of Aerobic Respiration and Facultative
Anaerobic Respiration to Degrade cBOD

Metabolic Function	Aerobic Respiration	Anoxic Respiration
Carbon substrate	cBOD	cBOD
Energy substrate	cBOD	cBOD
Final electron acceptor	O_2	NO_3^-
Gaseous end products	CO_2	CO_2, N_2O, N_2
Energy yield (per mole)	686 kcal	636 kcal
Sludge yield (per lb. cBOD)	0.6	0.4

Denitrifying bacteria are ubiquitous and are found in soil, water, and fecal waste. Commonly referenced genera of bacteria that have denitrifying species are *Bacillus* and *Pseudomonas*. The most widespread in wastewater are *Pseudomonas fluorescens* and *Pseudomonas denitrificans*. Other genera of bacteria that have denitrifying bacteria include *Acinetobacter, Agrobacterium, Alcaligenes, Corynebacterium, Hyphomicrobium, Propionobacterium, Rhizobium, Sprillium,* and *Thiobacillus*. Approximately 80% of the bacterial population in mixed liquor consists of facultative anaerobic bacteria. Nitrifying bacteria make up approximately 10% of the bacterial population in mixed liquor.

Denitrifying bacteria enter the activated sludge process through fecal waste and inflow and infiltration (I/I) as soil and water organisms. The generation time for most denitrifying bacteria is 20 to 30 minutes. As temperature increases, the growth rate of denitrifying bacteria also increases. Many denitrifying bacteria are floc-forming organisms, and some denitrifying bacteria such as *Microthrix parvicella* grow as filaments.

Microthrix parvicella and *Escherichia coli* are two unique denitrifying bacteria. They use nitrate to degrade cBOD, but only reduce nitrate to nitrite (NO_2^-). An additional denitrifying bacterium must then reduce nitrite to molecular nitrogen. Therefore, in addition to operational conditions that can contribute to a build-up of intermediates during denitrification, there are denitrifying bacteria that also can contribute to a build-up of intermediates (Equation 8.1). These intermediates are toxic, and the build-up of nitrite produces a significant chlorine demand.

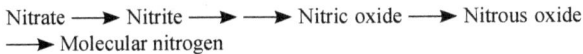

Nitrate $\longrightarrow$ Nitrite $\longrightarrow$ $\longrightarrow$ Nitric oxide $\longrightarrow$ Nitrous oxide $\longrightarrow$ Molecular nitrogen

$$NO_3^- \longrightarrow NO_2^- \longrightarrow NO \longrightarrow N_2O \longrightarrow N_2 \qquad \text{Equation 8.1}$$

If denitrifying bacteria are stressed by unfavorable operational conditions, their metabolism is incomplete, and intermediate compounds also may be produced and accumulate in relatively large quantities. The intermediate compounds include: 1) nitrite (NO_2^-), 2) nitric oxide (NO), and 3) nitrous oxide (N_2O). Denitrifying bacteria are sensitive to the carbon source used for denitrification, and the carbon source used may contribute to the production and accumulation of intermediates. Changes in carbon sources influence the dominant groups of denitrifying bacteria. Changes in carbon sources may temporarily result in a brief reduction in the rate of denitrification.

If denitrifying bacteria are needed, there are three operational measures that can be used to increase their numbers. Mixed liquor from another activated sludge process can be added. Bio-augmentation cultures of denitrifying bacteria may be introduced, or the process may be allowed to simply increase the number of denitrifying bacteria through natural seeding as the bacteria enter the treatment process over time through fecal waste and I/I.

A relatively small population of denitrifying bacteria or a population of sluggish denitrifying bacteria may occur during the following operational conditions: 1) start-up, 2) excess hydraulic loading from I/I, 3) overwasting, 4) inhibition or toxicity, and 5) recovery from toxicity. Except for substrate toxicity and recognizable cBOD toxicity, if toxicity does not occur for nitrifying bacteria, toxicity should not occur for denitrification.

Chapter 9

Denitrification and Limiting Factors

Denitrification is the removal of nitrogen from wastewater. Removal occurs when nitrogen is stripped from nitrate and escapes to the atmosphere as insoluble molecular nitrogen gas (N_2) and nitric oxide (N_2O). Nitrate is produced through nitrification. When nitrification and denitrification are coupled, nitrogen is removed from the wastewater. The effects of nitrification and denitrification upon the activated sludge process are significantly different (Table 9.1).

Table 9.1 Effects of Nitrification and Denitrification upon the Activated Sludge Process

Parameter	Nitrification	Denitrification
Molecule used for respiration	O_2	NO_3^-
Oxygen consumed per lb. NH_4^+ oxidized to NO_3^-	4.2 lbs.	---
Carbon as methanol (CH_3OH) oxidized per lb. NO_3^- removed	---	2.9 lbs.
Alkalinity consumed per lb. NH_4^+ oxidized to NO_3^-	7.1 lbs.	---
Alkalinity produced per lb. NO_3^- consumed	---	3.6
pH change	Decrease	Increase
Sludge produced per lb. NH_4^+ oxidized to NO_3^-	0.12.	---
Sludge produced per lb. carbon as methanol (CH_3OH) oxidized		0.45 lb.

Nitrate

Nitrate contamination is a major problem affecting water quality and public health. Nitrate in surface water can be used directly by aquatic plants and algae as their nitrogen nutrient for growth. This growth contributes to eutrophication and deterioration of water sources. Nitrate contamination is a major problem in groundwater. The consumption of excess quantities of nitrate (> 10 mg/L) in water used for drinking can cause abortions, abdominal pain, birth defects, cancer, hypertension, muscular weakness, and methemoglobinemia (blue baby syndrome).

Because nitrate is stable and highly soluble in wastewater, it is difficult to remove nitrogen from nitrate by conventional treatment processes such as coagulation, filtration, and precipitation. Therefore, it is biologically removed by nitrification and denitrification (Figure 9.1). Because nitrogen is not incorporated in cellular material during denitrification, denitrification is dissimilatory nitrate reduction (Figure 9.2).

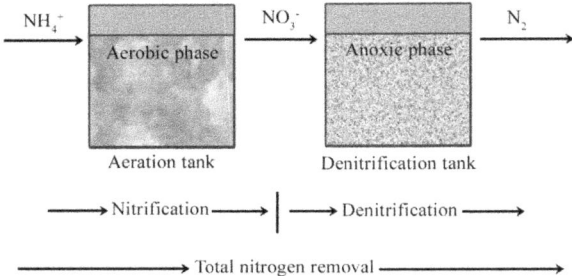

Figure 9.1 Sequential nitrification/denitrification: Because nitrate is highly soluble in wastewater, it is difficult to remove nitrogen from the wastewater. Therefore, nitrogen in the form of ammonium must first be nitrified to nitrate and then denitrified to nitrogenous gases, molecular nitrogen, and nitrous oxide. This is accomplished through sequential nitrification and denitrification. Sequential nitrification/denitrification can be used for total nitrogen removal.

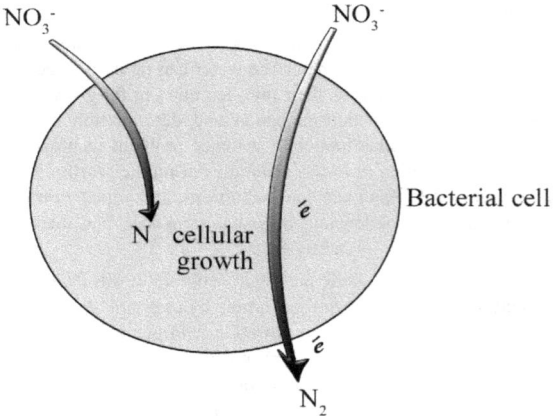

Figure 9.2 Assimilatory and dissimilatory nitrate reduction: There are two bacterial activities that reduce nitrate (remove oxygen from nitrogen) in the activated sludge process: assimilatory and dissimilatory nitrate reduction. In the absence of ammonium as a nitrogen nutrient for cellular growth, bacteria absorb nitrate and use nitrate as their nutrient source for growth. Here, nitrate is reduced, and the nitrogen is incorporated or assimilated into cellular material. Nitrogen does not leave the cell. When dissolved oxygen is no longer available for respiration, bacteria absorb nitrate and use the nitrate as an electron transport molecule to remove from the cell the electrons freed from degrading soluble cBOD. Here, nitrogen leaves the cell. It is not assimilated into cellular material. Dissimilatory nitrate reduction occurs.

When nitrate is used by denitrifying bacteria for cellular respiration, oxygen is not produced or returned to the treatment process. Theoretically, when nitrate is used for respiration each nitrate unit represents an oxygen equivalent of 2.86 units returned to the treatment process. Each unit of nitrate oxidizes approximately 4.0 units of cBOD.

Limiting Factors

Denitrification in the activated sludge process is influenced by four major limiting factors. The factors include:

- nitrate (or nitrite)
- adequate and active population of denitrifying bacteria
- absence of oxygen or an oxygen gradient
- presence of a carbon source (simplistic soluble cBOD)

Presence of nitrate (or nitrite)

Nitrate is produced in the activated sludge process through nitrification. Nitrate may enter the process through an industrial discharge, and it may be added to the sewer system as calcium nitrate $(Ca(NO_3)_2)$ or sodium nitrate $(NaNO_3)$ to control odors. Nitrite is produced in the activated sludge process through incomplete nitrification, and it may enter the process through an industrial discharge.

Except for nitrate addition for odor control, the presence of nitrate or nitrite in the sewer system is problematic. Existing conditions in the sewer system include little dissolved oxygen or the lack of dissolved oxygen and the absence of nitrate or nitrite. Sulfate (SO_4^{2-}) reduction does occur in the sewer system as well as mixed acid production (fermentation) and some methane production (Figure 9.3). Sulfate in groundwater enters the sewer system through inflow and infiltration (I/I) and as a component of urine.

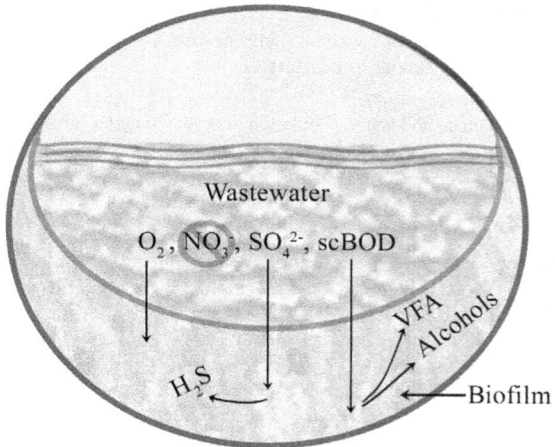

Figure 9.3 Bacterial respiration and fermentation in the sewer system: Electron transport molecules available in the sewer system for respiration consist of a relatively small quantity of dissolved oxygen and sulfate. Because conditions in the sewer system inhibit nitrification, nitrate is not available for respiration, unless an industry discharges nitrate to the sewer system. Soluble cBOD is available for fermentation. Therefore, bacteria in the biofilm and sediment use dissolved oxygen if it is available. Once the dissolved oxygen is depleted, sulfate-reducing bacteria use sulfate for respiration, resulting in the production of hydrogen sulfide that inhibits nitrification. Once dissolved oxygen is depleted, fermentative bacteria ferment soluble cBOD, resulting in the production of a mixture of acids and alcohols. Some of the acids and alcohols inhibit nitrification.

Bacterial respiration is the use of an inorganic molecule such as oxygen (O_2), nitrate (NO_3^-), or sulfate (SO_4^{2-}) to remove electrons released during substrate degradation. Fermentation is the use of an organic molecule to remove electrons released during substrate degradation. (Figure 9.4).

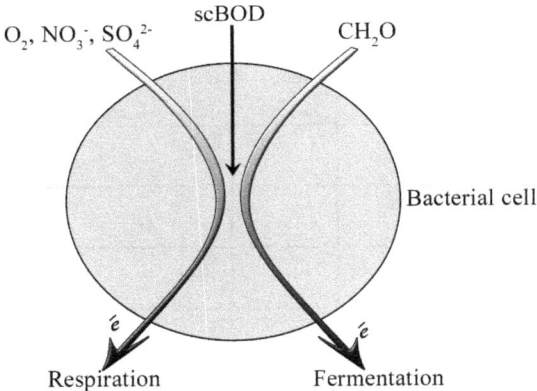

Figure 9.4 Respiration and fermentation: Respiration is the use of an inorganic molecule to degrade substrate. Inorganic molecules used for respiration include free molecular (dissolved) oxygen, nitrate, and sulfate. Fermentation is the use of an organic molecule to degrade substrate. Organic molecules used for fermentation consist of soluble cBOD. The purpose of these molecules, inorganic and organic, is to remove electrons released from the degrading substrate.

Bacterial respiration via sulfate reduction produces sulfides (HS^-) and hydrogen sulfide (H_2S) and a relatively small quantity of cells (sludge or biofilm) (Table 9.2). Mixed acid production also produces odors in the form of volatile organic compounds (VOC) including volatile fatty acids (VFA) and a relatively small quantity of cells (sludge or biofilm) (Table 9.3). Methane production in the sewer system occurs, but it yields only a relatively small quantity of cells (sludge or biofilm) (Table 9.1). However, the use of oxygen or nitrate for respiration produces a relatively large quantity of cells (sludge or biofilm).

Table 9.2 Respiration and Sludge Yield

Form of Oxidation of Substrate	Molecule Used	Approximate Sludge Yield (lbs. sludge per lb. cBOD degraded)
Aerobic/oxic (respiration)	O_2	0.6
Anaerobic/anoxic (respiration)	NO_3^-	0.4
Anaerobic/sulfate reduction (respiration)	SO_4^{2-}	0.1
Anaerobic/mixed acid production (fermentation)	CH_2O*	0.1
Anaerobic/methane production (fermentation)	CH_2O*	0.05

*CH_2O designates an organic molecule.

By introducing nitrate to the sewer system, denitrifying bacteria use the nitrate to degrade soluble cBOD. This results in increased cellular production or biofilm production (Figure 9.5a). The increase in biofilm reduces the diameter of the sewer main through which wastewater is conveyed to the treatment plant. By degrading a large quantity of soluble cBOD in the sewer system, less cBOD enters the treatment plant. This reduction in influent cBOD may result in the inability of the treatment process to satisfy 85% removal efficiency for cBOD or BOD as required by permit. Additional problematic conditions that can occur in the treatment plant include primary clarifier bulking, toxicity to the biomass, chlorine sponge, thickener bulking, and an increase in anaerobic digester ORP (oxidation-reduction potential) and anaerobic digester foaming, and a decrease in methane production (Figure 9.5b).

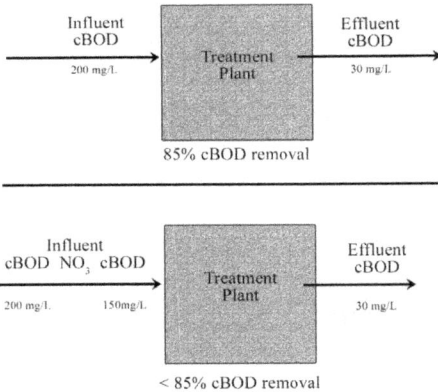

Figure 9.5a Problematic conditions caused by bacterial respiration using nitrate in the sewer system: Biological wastewater treatment plants typically have an 85% reduction requirement for cBOD or BOD across the treatment process. For example, if the cBOD influent is 200 mg/L, then the cBOD effluent must be $\leq$ 30 mg/L to achieve at least an 85% reduction in cBOD. The same treatment plant typically has a discharge concentration requirement for BOD or cBOD. This requirement usually is $\leq$ 30 mg/L. Without nitrate in the sewer system, little cBOD is reduced in the sewer system. Therefore, if the treatment plant discharges 30 mg/L cBOD and has an influent 200 mg/L, the plant has satisfied both requirements: 85% reduction in cBOD across the plant and the discharge concentration of $\leq$ 30 mg/L (top schematic). However, if nitrate is discharged to the sewer system or accepted for discharge in the sewer system, bacteria in the sediment and biofilm will use the nitrate to degrade cBOD. This degradation of cBOD in the sewer system results in decreased cBOD entering the treatment plant. If the influent cBOD is usually 200 mg/L and 50 mg/L cBOD are degraded in the sewer system due to the presence of nitrate, only 150 mg/L cBOD is received at the treatment plant. If the effluent cBOD is 30 mg/L, an 85% reduction in cBOD across the treatment plant does not occur even though the effluent concentration of cBOD complies with the discharge requirement (bottom schematic). Also, the degradation of cBOD in the sewer system with nitrate results in an increase in the biofilm, resulting in a decrease in the diameter of the sewer main for the conveyance of wastewater and possible odor production.

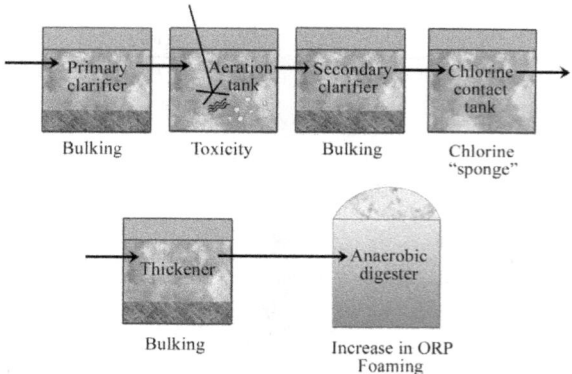

Figure 9.5b Problematic conditions in the treatment plant caused by influent nitrite and nitrate: If nitrate or nitrite enters the treatment plant and the activated sludge process, there are several operational problems that may occur: 1) denitrification and floating solids in the primary clarifier, 2) inhibition or toxicity in the activated sludge process, 3) pass-through of nitrate to the secondary clarifier due to the presence of dissolved oxygen in the aeration tank, 3) denitrification in the secondary clarifier, 4) interference with chlorine disinfection of the effluent due to the "chlorine sponge," 5) denitrification of the thickener, resulting in excess loss of solids from the thickener to the head of the treatment plant, 6) foam production in the anaerobic digester as molecular nitrogen and nitrous oxide produced through denitrification are entrapped in solids that rise in the annular space of the digester, and 7) an increase in ORP in the anaerobic digester, resulting in decreased methane production.

There are two nitrate-containing compounds, sodium nitrate ($NaNO_3$) and calcium nitrate ($Ca(NO_3)_2$), that are added to the sewer system to control odors including hydrogen sulfide and volatile fatty acids. By adding sodium nitrate or calcium nitrate there is an increase in ORP, which inhibits sulfate reduction and mixed acid production (Table 9.3), and the use of nitrate by denitrifying bacteria for respiration produces nitric oxide (NO) and nitrous oxide (N_2O), which are inhibitory to sulfate-reducing bacteria (SRB).

Table 9.3 Guideline ORP Values and Cellular Activity

Cellular Activity	ORP, mV
Nitrification	+100 to +350
cBOD degradation with oxygen	+50 to +250
Biological phosphorus uptake	+25 to +250
Denitrification	+50 to −50
Sulfide production (sulfate reduction)	−50 to −250
Biological phosphorus release	−100 to −250
Acid formation (mixed acid production/fermentation)	−100 to −225
Methane production ($CO_2 + H_2$)	−175
Methane production (degradation of organic compounds)	−300

ORP is monitored as millivolts (mV). It measures the quantity of oxidized ions and compounds in the wastewater or sludge. With mostly oxidized ions and compounds in the wastewater or sludge the ORP is positive. ORP also measures the quantity of reduced organic compounds in the wastewater or sludge. With mostly reduced compounds in the wastewater or sludge the ORP is negative. The more oxidized the wastewater or sludge, the higher the positive value. The more reduced the wastewater or sludge, the lower the negative value. An ORP value of zero is a balance of oxidized and reduced components in the wastewater, just as a pH of 7.0 is a balance of acidic and alkali components in the wastewater.

Absence of oxygen or an oxygen gradient

Denitrifying bacteria have the ability to use either free molecular oxygen or nitrate to degrade cBOD. However, the bacteria can use only one molecule at a time. Therefore, if free molecular oxygen and nitrate are both available, denitrifying bacteria will use free molecular oxygen. The use of free molecular oxygen over nitrate is that respiration with free molecular oxygen provides more energy and more growth (sludge yield) than respiration with nitrate (Table 9.4). In the presence of free molecular oxygen, the oxygen stops or "inhibits" the use of nitrate by suppressing the nitrate-reducing enzyme production in facultative anaerobic heterotrophs.

Table 9.4 Comparison of Energy Obtained and Sludge Yield When Denitrifying Bacteria Use Oxygen or Nitrate

Metabolic Function	Aerobic Respiration	Anoxic Respiration
Carbon substrate	cBOD	cBOD
Energy substrate	cBOD	cBOD
Molecule used for respiration	O_2	NO_3^-
Energy yield (per mole)	686 kcal	636 kcal
Sludge yield (per lb. cBOD)	0.6	0.4

The use of free molecular oxygen and nitrate can occur simultaneously, if an oxygen gradient is present (Figure 9.6). An oxygen gradient is present in a floc particle (solids) if: 1) the floc particle is > 100 μm in diameter, 2) the dissolved oxygen concentration < 1 mg/L in the bulk solution, 3) soluble cBOD is present in the bulk solution, and 4) nitrate is present in the bulk solution. Denitrifying bacteria are present in the perimeter of the floc particle as well as in the core of the floc particle. The bacteria in the perimeter are exposed to free molecular oxygen and nitrate, but only use free molecular oxygen. As oxygen, nitrate, and soluble cBOD penetrate the floc particle oxygen will be used; nitrate will not be used as long as oxygen is present. Once oxygen is depleted near the core of the floc particle, the denitrifying bacteria in the core will then use nitrate. Therefore, in the presence of free molecular oxygen (< 1 mg/L) denitrification can occur, if an oxygen gradient is present.

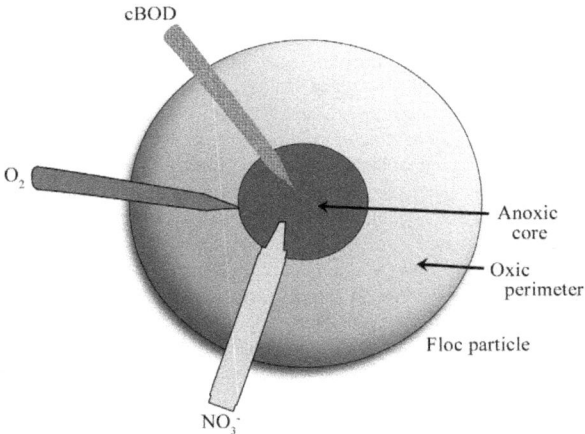

Figure 9.6 Simultaneous use of oxygen and nitrate by heterotrophic bacteria: In the presence of an oxygen gradient where dissolved oxygen and nitrate are available for bacterial respiration, both oxygen and nitrate can be used simultaneously if: 1) soluble cBOD penetrates into the core of the floc particle, 2) nitrate penetrates to the core of the floc particle, and 3) dissolved oxygen is relatively small in concentration (< 1.0 mg/L). This condition would produce an oxic (aerobic) perimeter where dissolved oxygen is used for respiration and an anoxic (absence of oxygen and presence of nitrate) core where nitrate is used for respiration.

Presence of a carbon source (simplistic soluble cBOD)

Carbon is the most critical limiting factor for denitrification and nitrification. The presence of carbon or soluble cBOD is the primary determinant of denitrification rates in wastewater. Carbon is used for growth (sludge production). If carbon is not available, growth cannot occur. If growth cannot occur, then energy is not needed. If energy is not needed, then the oxidation of ammonium and nitrite (nitrification) does not occur, and the oxidation of cBOD (denitrification) does not occur.

However, carbon needed for denitrification must be soluble. Only soluble cBOD can enter a bacterial cell (Figure 9.7). Highly soluble, simplistic compounds such as acetic acid (CH_3COOH) (Equation 9.1), ethanol (CH_3CH_2OH) (Equation 9.2), methanol (CH_3OH) (Equation 9.3), glucose ($C_6H_{12}O_6$), and sodium acetate (CH_3COONa), all of which can be easily assimilated by denitrifying bacteria, are excellent carbon substrates (Table 9.5). Approximately 2.9 lbs. methanol is consumed per lb. nitrate-nitrogen denitrified. This is equivalent to 2.86 lbs. COD degraded per lb. nitrate-nitrogen degraded. Atypical compounds or materials used as a carbon include: glycerol, lactic acid, molasses, peptone, and sawdust. Commercially prepared carbon sources also are available. The type of carbon used for denitrification has a significant influence on not only the denitrification rate but also sludge yield and diversity of the bacterial population.

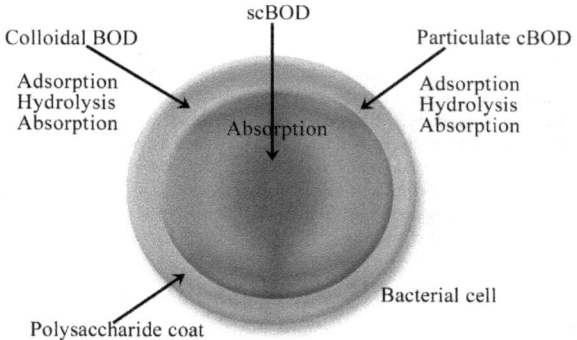

Figure 9.7 Absorption and adsorption of cBOD: Bacteria can only degrade soluble cBOD. In order for cBOD degradation to occur, it must enter the bacterial cell through absorption. Soluble cBOD is absorbed by bacteria and degraded intracellularly. Colloidal BOD such as proteins and particulate BOD such as starches cannot be absorbed by bacteria. They require time to be degraded. They first must be adsorbed to the surface of the bacterial cell and then undergo hydrolysis where the colloids and particulates are converted to soluble simplistic forms of cBOD such as short-chain acids, alcohols, and

amines. These soluble forms of cBOD can then be absorbed by the bacteria and degraded intracellularly. Therefore, only soluble cBOD can promote rapid denitrification.

Acetic acid + Nitrate —— Denitrifying bacteria ——→

Carbon dioxide + Molecular nitrogen + Water + Hydroxyl

$$5\ CH_3COOH + 8\ NO_3^- \text{—— Denitrifying bacteria ——→}$$

$$10\ CO_2 + 4\ N_2 + 7\ H_2O + 8\ OH^- \qquad \text{Equation 9.1}$$

Ethanol + Nitrate —— Denitrifying bacteria ——→

Carbon dioxide + Molecular nitrogen + Water + Hydroxyl

$$5\ CH_3CH_2OH + 12\ NO_3^- \text{—— Denitrifying bacteria ——→}$$

$$10\ CO_2 + 6\ N_2 + 9\ H_2O + 12\ OH^- \qquad \text{Equation 9.2}$$

Nitrate + Methanol —— Denitrifying bacteria ——→

Carbon dioxide + Molecular nitrogen + Water + Hydroxyl

$$6\ NO_3^- + 5\ CH_3OH \text{—— Denitrifying bacteria ——→}$$

$$5\ CO_2 + 3\ N_2 + 7\ H_2O + 6\ OH^- \qquad \text{Equation 9.3}$$

Table 9.5 Examples of Highly Soluble, Easily Assimilated Carbon Sources for Denitrification

Carbon Source	Formula	mg Carbon Source required per mg NO_3^-
Acetic acid	CH_3COOH	0.79
Ethanol	CH_3CH_2OH	0.45
Methanol	CH_3OH	0.29

The rate of denitrification varies greatly depending on the carbon source (cBOD) used by denitrifying bacteria. The highest rates are obtained with the addition of acetic acid and methanol. The presence and use of these compounds quickly removes any residual dissolved oxygen in the reactor and then permits rapid denitrification. Low rates of denitrification are obtained with raw wastewater or primary clarifier effluent as the carbon source, and the lowest rates of denitrification are obtained with endogenous decay as the carbon source (Figures 9.8). Because an organic compound is used by denitrifying bacteria for growth, it may be necessary to add phosphorus (orthophosphate) as a nutrient to the denitrification reaction.

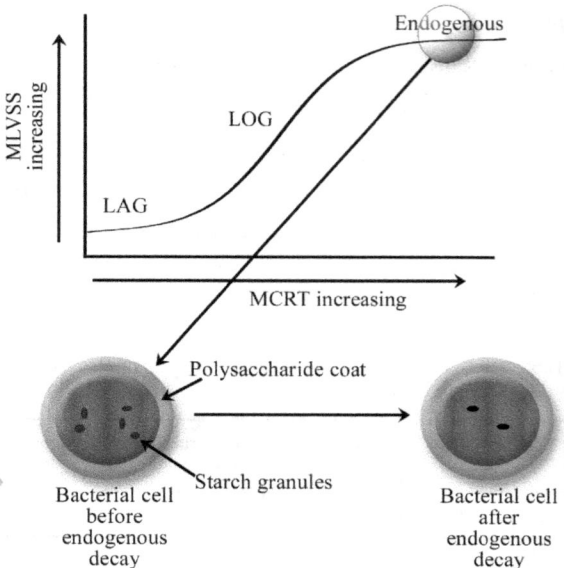

Figure 9.8 Endogenous decay: Endogenous decay occurs when cBOD in the bulk solution is not available or limited in quantity. When this occurs bacteria solubilize stored foods, polysaccharide coatings, and starch granules. On the bacterial growth curve endogenous respiration occurs when substrate (influent cBOD) is received in a daily steady-state loading condition. A steady-state condition is when the cBOD loading is within a range of cBOD loading values typically observed at the treatment process. During low loading conditions such as early-morning hours (1 AM to 5 AM), endogenous decay occurs.

The addition of methanol or other chemical compound as a carbon source for denitrification should be done carefully. Overdosing of a carbon source adds to the cost of treatment and may contribute to an increase in effluent cBOD. Insufficient carbon addition contributes to an increase in effluent nitrite and/or nitrite as well as total nitrogen.

Methanol or "wood alcohol" is the simplest alcohol and an excellent carbon (cBOD) source for denitrifying bacteria. It is easily assimilated, converting nitrate to molecular nitrogen, and it provides the highest nitrogen removal efficiency.

Methanol is classified as a 1B flammable liquid by the National Fire Protection Association (NFPA) and has an NFPA 704 hazard identification rating of 3 for flammability, 1 for health, and 0 for reactivity. Methanol is toxic to humans and the environment. Appropriate planning and training for the use of methanol is critical for wastewater treatment plant operators. Training should include: handling, storage, prevention of spills and vapors, and protection against fire and explosion.

pH

The optimum pH range for denitrifying bacteria is 7.0 to 9.0. Rates of denitrification decrease significantly at pH values outside this range. At lower pH values the production and accumulation of nitric oxide (NO) and nitrous oxide (N_2O) occurs.

Temperature

In addition to substrate dependency, temperature dependency also occurs in denitrifying bacteria. With increasing temperature the rate of denitrification increases. Denitrifying bacteria have a wider range of temperature values in which they are active as compared to nitrifying bacteria.

Temperature $< 20°C$ significantly affects denitrification. To compensate for depressed temperature in temperate regions, longer hydraulic retention time (HRT) may be needed as well as an increase in MLVSS. Denitrification rates at 10°C can be 20 to 40% lower than rates at 20°C. The best range of temperatures for denitrification is 20 to 25°C. Some depression in the rate of denitrification can occur at temperatures $> 25°C$ (Figure 9.9).

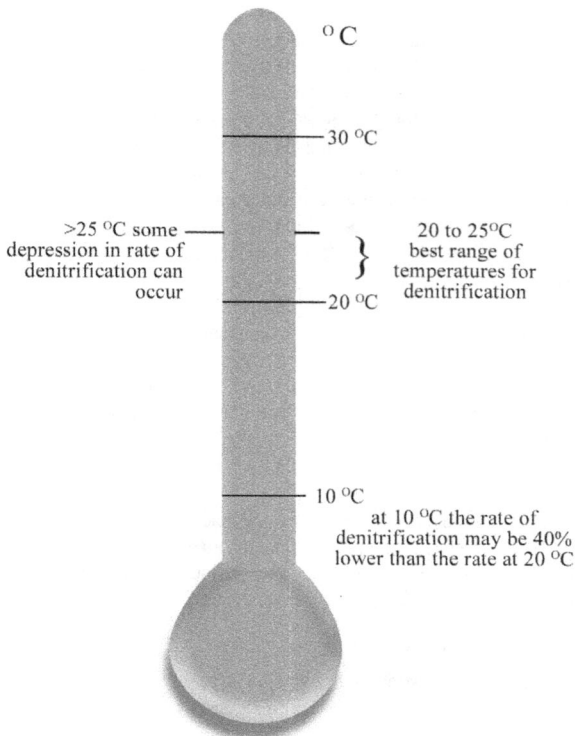

°C

— 30 °C

>25 °C some —
depression in rate of
denitrification can
occur

} 20 to 25°C
best range of
temperatures for
denitrification

— 20 °C

— 10 °C

at 10 °C the rate of
denitrification may be 40%
lower than the rate at 20 °C

Figure 9.9 Temperature and its impact upon denitrification: Temperature has a profound impact upon the population size of the denitrifying bacteria and the activity of the denitrifying bacteria. Increasing temperature favors the growth of denitrifying bacteria and promotes the activity of the bacteria. Decreasing temperature hinders the growth of denitrifying bacteria and decreases the activity of the bacteria. There is one critical temperature and one critical temperature range. The temperature value of concern is 10°C when the rate of denitrification is approximately 40% of the rate of denitrification at 20°C. The temperature range is 20 to 25°C when the rate of denitrification is best.

Inhibition and toxicity

Because denitrifying bacteria obtain more energy from the oxidation of cBOD than nitrifying bacteria obtain from the oxidation of an equivalent amount of nBOD, denitrifying bacteria are much more resilient than nitrifying bacteria to inhibition and toxicity. This difference in energy enables denitrifying bacteria to acclimate more quickly and more often to inhibition than nitrifying bacteria. This difference in energy also enables denitrifying bacteria to reproduce more quickly than nitrifying bacteria.

Removal of nitrogen by nitrification and denitrification can be performed using several modes of operation. Removal can be achieved using a single-unit process having various treatment zones such as the Modified Ludzack-Ettinger process (Figure 9.10) or in separate stages such as the Four-stage Bardenpho process (Figure 9.11).

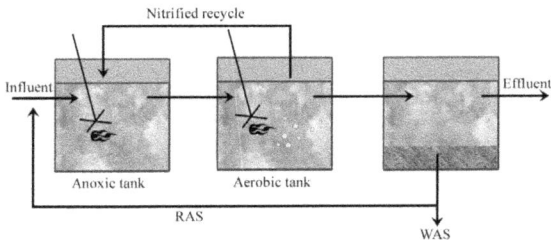

Figure 9.10 Modified Ludzack-Ettinger process: The Modified Ludzack-Ettinger process is a mainstream process where nutrient removal for nitrogen occurs in plug-flow mode of operation. There is no side stream to a reactor not in the plug-flow mode. In the Modified Ludzack-Ettinger process denitrification occurs in the first reactor (anoxic tank) where nitrate from an aeration tank is recycled to the anoxic selector. Nitrification occurs in the aeration tank.

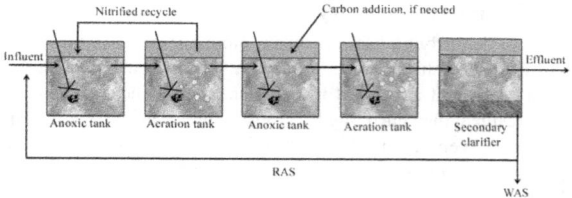

Figure 9.11 Four-stage Bardenpho process: The four-stage Bardenpho process is a mainstream process where nutrient removal for nitrogen occurs in two anoxic (denitrification) tanks. Nitrification occurs in two aeration tanks and nitrified mixed liquor is recycled to the anoxic tanks. If needed, a carbon source such as methanol may be added to an anoxic tank.

PART THREE: BIOLOGICAL PHOSPHORUS REMOVAL

Chapter 10

Biological Phosphorus Removal: The Basics

There are four basic operational measures that can be used to re-move phosphorus from the activated sludge process: 1) chemical precipitation, 2) bacterial assimilation, 3) enhanced biological phos-phorus removal (EBPR) or luxury uptake of phosphorus by Poly-P bacteria, and 4) a combination of EBPR and chemical precipitation and/or tertiary treatment (Table 10.1).

Table 10.1 Operational Measures Available for Phosphorus Removal

Operational Measure	Comments
Chemical precipitation	Use of aluminum and iron salts, lime, or polymer to precipitate phosphate
Bacterial assimilation	Uptake and incorporation of phosphorus in cellular material or MLVSS and wasting of MLVSS. Phosphorus makes up 1 to 3% of dry weight of bacterial cell or MLVSS.
Luxury uptake of phosphorus	Uptake and incorporation of phosphorus in cellular material or MLVSS and production of polyphosphate or volutin granules by Poly-P bacteria. Phosphorus makes up 6 to 7% of dry weight of bacterial cell or MLVSS.
Luxury uptake of phosphorus and chemical precipitation and/or tertiary treatment	Release of phosphate by Poly-P bacteria in an anaerobic reactor and chemical precipitation of phosphate and/or cloth or sand filtration.

Chemical precipitation of phosphorus

In order to chemically precipitate phosphorus from wastewater, phosphorus must be in the phosphate form. Phosphate is soluble and is the only form of phosphorus that reacts with a precipitant or polymer. Because phosphate reacts with a precipitant or polymer it is known as "reactive" phosphorus.

A variety of chemicals are used to precipitate phosphate. The coagulant used may be applied as a one-point chemical addition (primary clarifier only) or two-point chemical addition (primary clarifier and secondary clarifier). Most commonly used chemicals include aluminum salts and iron salts (Table 10.2), lime, and polymers. Chemical precipitation of phosphate results in the production of solids, and the use of aluminum and iron salts results in the consuming of alkalinity and reduction of pH.

Table 10.2 Aluminum and Iron Salts Used to Precipitate Phosphate

Metal Salt	Formula
Aluminum chloride	$AlCl_3$
Aluminum chlorohydrate	$Al_nCl(3_{n-m})(OH)_m$*
Aluminum sulfate	$Al_2(SO_4)_3$
Poly-aluminum chloride	$Al_n(OH)_mCl_{3n-m}$*
Sodium aluminate	$Na_2Al_2O_4$
Ferric chloride	$FeCl_3$
Ferrous chloride	$FeCl_2$
Ferrous sulfate	$FeSO_4$

*Aluminum chlorohydrate is a group of aluminum salts having the formula as listed. Poly-aluminum chloride is a group of aluminum salts having the formula as listed.

Aluminum salts are available for use as aluminum chloride, aluminum chlorohydrate, aluminum sulfate or alum, sodium aluminate, and poly-aluminum chloride or PAC. Approximately 0.87 pounds of aluminum are required to precipitate one pound of phosphorus as P. With low total phosphorus discharge limits (< 1.0 mg/L), the quantity of aluminum needed to precipitate one pound of phosphorus increases. Phosphate is precipitated as aluminum phosphate (Equation 10.1), and alkalinity is consumed (Equation 10.2).

Aluminum + Phosphate $\longrightarrow$ Aluminum phosphate

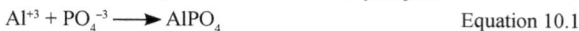

$$Al^{+3} + PO_4^{-3} \longrightarrow AlPO_4 \qquad \text{Equation 10.1}$$

Aluminum + Bicarbonate alkalinity $\longrightarrow$ Aluminum hydroxide

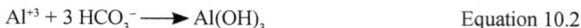

$$Al^{+3} + 3\ HCO_3^- \longrightarrow Al(OH)_3 \qquad \text{Equation 10.2}$$

Iron salts are available for use as ferric chloride, ferrous chloride, and ferrous sulfate. Approximately 1.8 pounds of iron are required to precipitate one pound of phosphorus as P (Equation 10.3). Like aluminum precipitation of phosphate, low total phosphorus discharge limits (< 1.0 mg/L) require an increase in the quantity of iron needed to precipitate one pound of phosphorus. Phosphate is precipitated as iron phosphate, and alkalinity is consumed (Equation 10.3), and (Equation 10.4).

Iron + Phosphate $\longrightarrow$ Iron phosphate

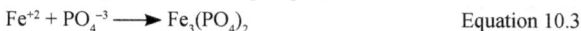

$$Fe^{+2} + PO_4^{-3} \longrightarrow Fe_3(PO_4)_2 \qquad \text{Equation 10.3}$$

Iron + Bicarbonate alkalinity $\longrightarrow$ Ferrous hydroxide

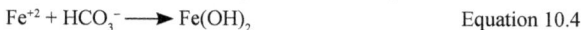

$$Fe^{+2} + HCO_3^- \longrightarrow Fe(OH)_2 \qquad \text{Equation 10.4}$$

With aluminum and iron salts the metals react with bicarbonate in the wastewater. Bicarbonate is found in municipal wastewater and is produced during cellular respiration when carbon dioxide is released. Carbon dioxide dissolves in the wastewater to produce carbonic acid (H_2CO_3). When carbonic acid dissociates in the wastewater, bicarbonate is produced (Equation 10.5).

Carbonic acid ⬌ Hydrogen proton + Bicarbonate

$$H_2CO_3 \longleftrightarrow H^+ + HCO_3^- \qquad \text{Equation 10.5}$$

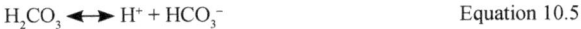

The quantity of lime as calcium hydroxide ($Ca(OH)_2$) or hydrated lime (slack lime) needed to precipitate phosphate is influenced by alkalinity. Approximately 1.5 pounds of lime are required per pound of alkalinity. In order to precipitate the phosphate the pH of the reactor must be 10.5. Precipitated phosphorus exists as a component of hydroxyapatite or calcium hydroxide phosphate ($Ca_5(OH)(PO_4)_3$).

Enhanced biological phosphorus removal (EBPR)

Enhanced biological phosphorus removal (EBPR) or luxury uptake of phosphorus and nitrogen removal may be performed by one of several mainstream or side stream (Figure 10.1) nutrient removal processes (Figures 10.2a, 10.2b): 1) A/O, 2) A^2/O, 3) Bardenpho, 4) Phostrip, and 5) University of Cape Town or UCT (Table 10.3). For biological phosphorus removal, each process has an upstream anaerobic/fermentation reactor and a downstream aerobic/oxic reactor (plug-flow mode of operation). For nitrogen removal, each process has an aerobic reactor for nitrification and a downstream or upstream anoxic reactor for denitrification.

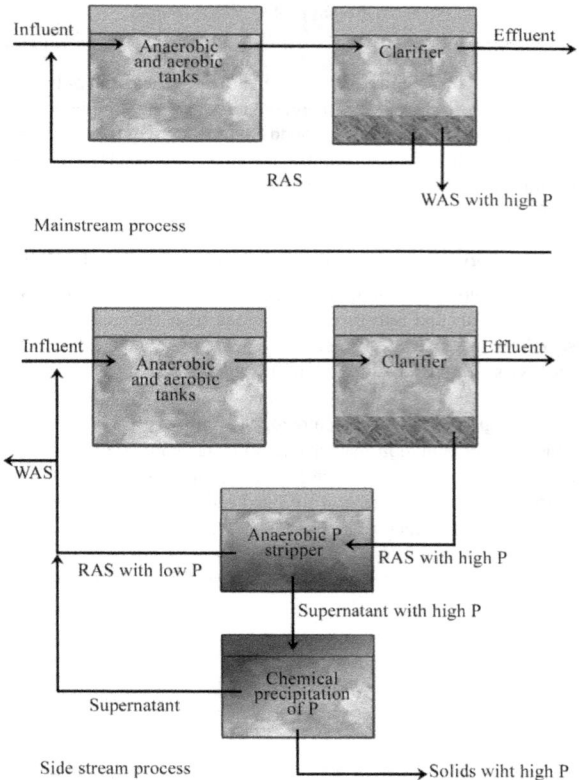

Figure 10.1 Mainstream and side steam processes for the removal of phosphorus: Biological nutrient removal can occur in mainstream and side stream processes. In mainstream processes the removal of nitrogen and/or phosphorus occurs in reactors that are aligned in plug-flow mode of operation. In side stream processes the removal of nitrogen and/or phosphorus occurs in reactors that are and are not aligned in plug-flow mode of operation.

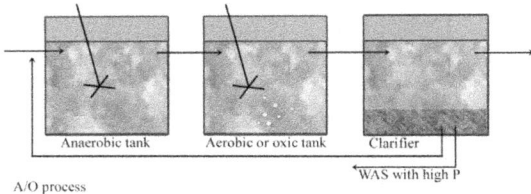

A/O process

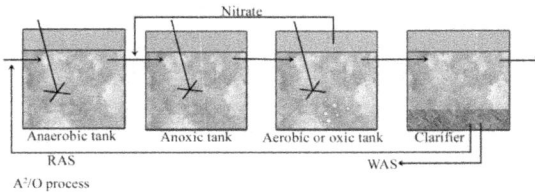

A²/O process

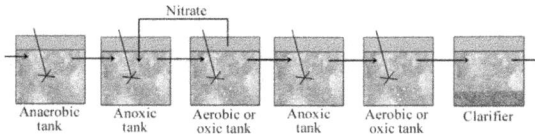

Modified Bardenpho process (5-stage Bardenpho process)

Figure 10.2a Nutrient removal processes for the removal of phosphorus or nitrogen: The A/O process consists of an anaerobic reactor upstream of an aerobic reactor. The anaerobic reactor acts as an anaerobic/fermentative reactor for biological phosphorus release, while the aerobic reactor performs biological phosphorus uptake. The A²/O is a mainstream process that removes phosphorus and nitrogen. The first reactor is an anaerobic/fermentative reactor for biological phosphorus release. The second reactor is an anoxic reactor for denitrification. Nitrified wastewater produced in the third or aerobic reactor is recycled to the anoxic reactor. The aerobic reactor also performs biological phosphorus uptake. The Modified Bardenpho process (five-stage Bardenpho process) is also a mainstream process for the removal of phosphorus and nitrogen.

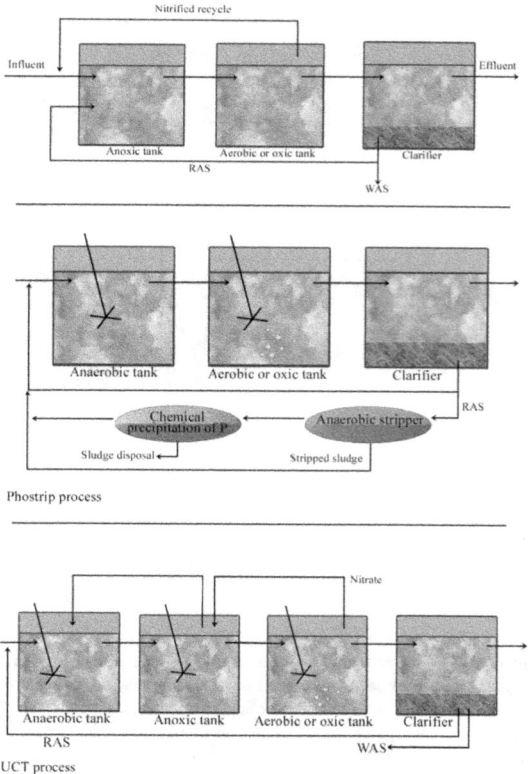

Figure 10.2b Nutrient removal processes for the removal of phosphorus or nitrogen: In the top schematic, biological phosphorus removal occurs in a mainstream process. In the middle schematic, the mainstream process is modified and operates as a side stream process. In this process secondary solids that contain a large number of Poly-P bacteria having large quantities of phosphorus are transferred to an anaerobic stripper. In the stripper the Poly-P bacteria release phosphorus to the bulk solution. After biological phosphorus

release, the settled solids are either wasted from the treatment process or return to the anaerobic reactor. The released phosphorus in the stripper is chemically precipitated to an inorganic sludge that is removed from the treatment process. The UCT process (bottom schematic) is capable of removing phosphorus and nitrogen. Nitrified wastewater from the aeration tank is recycled to the anoxic tank where denitrification occurs, and solids from the anoxic tank are recycled to the anaerobic tank to increase solids concentration to provide for an anaerobic/fermentative condition where biological phosphorus release occurs. Biological phosphorus uptake occurs in the aerobic tank.

Table 10.3 Nutrient Removal Processes

Process	Nutrient Removed		Mainstream or Side Stream		Chemical Precipitation of Phosphorus
	Nitrogen	Phosphorus	Mainstream	Side Stream	
A/O		X	X		
A²/O	X	X	X		
Bardenpho	X	X	X		
Modified Ludzack-Ettinger (MLE)	X				
Phostrip	X	X		X	X
UCT	X	X	X		

EBPR is the removal of phosphorus from wastewater by bacteria in quantities beyond their cellular needs and then the wasting of the phosphorus-laden bacteria (sludge) from the activated sludge process. EBPR is capable of removing 80 to 90% of the influent phosphorus and producing < 1 mg/L TP in the effluent. EBPR requires two reactors, two groups of heterotrophic bacteria, and two major biochemical events.

The reactors needed are an upstream, anaerobic/fermentative reactor and a downstream aerobic reactor. The bacteria needed are fermentative or acid-forming bacteria and phosphorus-accumulating or Poly-P bacteria. The two major biochemical events are biological phosphorus release and biological phosphorus uptake.

In the first reactor, biological phosphorus release occurs under an anaerobic condition. In the second reactor, biological phosphorus uptake occurs under an aerobic condition. Fermentative bacteria (Table 10.4) and Poly-P bacteria (Table 10.5) are present in each reactor. The bacteria enter the activated sludge process through fecal waste and inflow and infiltration (I/I) as soil and water bacteria. Fermentative bacteria under an anaerobic condition produce volatile fatty acids including acetic acid and other short-chain, highly soluble organic compounds that serve as substrate for Poly-P bacteria (Table 10.6). Acetic acid is the most easily assimilated and is often added as a supplemental substrate to the anaerobic/fermentative reactor. Poly-P bacteria absorb these compounds under an anaerobic condition and then under an aerobic condition absorb phosphorus and store it in concentrations higher than cellular needs. In addition to Poly-P bacteria, other phosphorus-accumulating organisms (PAO) include algae, Cyanobacteria, fungi (yeast and filamentous forms), and protozoa.

Table 10.4 Fermentative or Acid-forming Bacteria

Acetobacterium
Citrobacter
Clostridium

| Enterobacter |
| Escherichia |
| Klebsiella |
| Lactococcus |
| Propionibacterium |
| Proteus |
| Salmonella |
| Serratia |

Table 10.5 Poly-P Bacteria

| Acineobacter |
| Aerobacter |
| Aeromonas |
| Arthrobacter |
| Beggiatoa |
| Enterobacter |
| Escherichia |
| Klebsiella |
| Moraxella |
| Mycobacterium |
| Pseudomonas |

Table 10.6 Volatile Fatty Acids

Fatty Acid	Chemical Formula
Formic acid	$HCOOH$
Acetic acid	CH_3COOH
Propionic acid	CH_3CH_2COOH

Fatty Acid	Chemical Formula
Butyric acid	$CH_3CH_2CH_2COOH$
Valeric acid	$CH_3CH_2CH_2CH_2COOH$
Caproic acid	$CH_3CH_2CH_2CH_2CH_2COOH$

Most Poly-P bacteria are aerobes. There are some Poly-P bacteria that can use nitrate (NO_3^-) for respiration and do absorb phosphate in the presence of nitrate and absence of oxygen. *Aeromonas* and *Pseudomonas* are the major Poly-P bacteria involved in EBPR as they represent more than 50% of the total aerobic Poly-P bacteria that participate in EBPR. The typical phosphorus content of bacteria is 1 to 3% (dry weight basis). Poly-P bacteria can accumulate up to 7 to 10% phosphorus (dry weight basis).

Biological phosphorus release (Figure 10.3)

In the anaerobic/fermentative reactor in the absence of free molecular oxygen and nitrate, fermentative bacteria produce the preferred substrates for Poly-P bacteria through fermentation of influent cBOD. Once produced, the Poly-P bacteria absorb the soluble substrate, but in the absence of free molecular oxygen they cannot degrade the substrate. However, the substrate is converted and stored as insoluble polyhydroxy-alkanoates (PHA). Polyhydroxy-alkanoates are stored intracellularly as granules that can be observed as red granules under methylene blue staining.

The conversion of soluble substrate to insoluble substrate as PHA requires energy. Bacteria store energy in high-energy phosphate bonds. When energy is needed for cellular metabolism including the conversion of soluble substrate to insoluble substrate, high-energy phosphate bonds are broken and phosphorus is released from the cell. Biological phosphorus release occurs. Therefore, as a result of biological phosphorus release there are two "pools" of phosphorus in the anaerobic/fermentative reactor: the phosphorus from biological phosphorus release and the influent phosphorus. Major biological activities in the anaerobic/fermentative reactor are: 1) fermentation and the production of soluble substrate for Poly-P bacteria and 2) release of phosphorus from Poly-P bacteria through the conversion of soluble substrate to insoluble PHA.

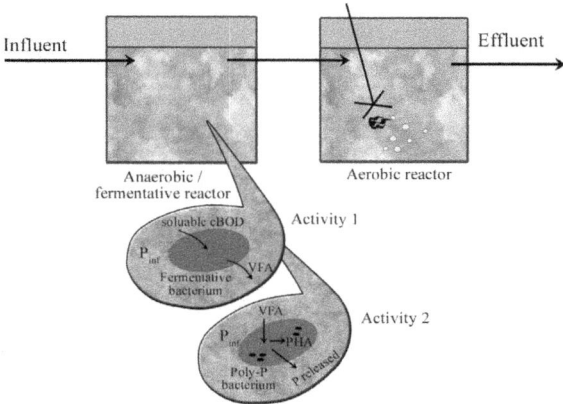

Figure 10.3 Biological phosphorus release: There are two critical biological events or activities that must occur in the anaerobic/fermentative reactor in order for biological phosphorus release to occur. First, acid-forming or fermentative bacteria must ferment influent soluble cBOD to simplistic volatile fatty acids (VFA) under an anaerobic condition. Second, after the production of VFA, Poly-P bacteria must absorb VFA and expend energy (biological phosphorus release) to convert the VFA to PHA and store the PHA.

Biological phosphorus uptake (Figure 10.4)

After biological phosphorus release, the Poly-P bacteria are carried by the wastewater to the downstream aerobic reactor. Here, in the presence of dissolved oxygen, PHA granules are solubilized and degraded by the bacteria. The degradation results in the release of a copious quantity of energy that is captured by the bacteria and stored as high-energy phosphate bonds. In order to store the energy, Poly-P bacteria absorb the released phosphorus and influent phosphorus. Excess phosphorus absorbed beyond cellular is stored as long chains or polymers of insoluble phosphate or polyphosphate. Polyphosphate is stored intracellularly as insoluble volutin granules.

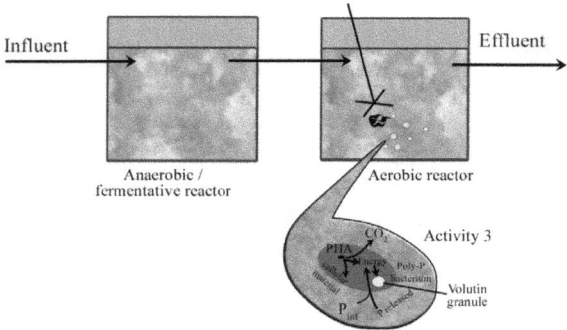

Figure 10.4 Biological phosphorus uptake: There is one critical biological event or activity that must occur in the aerobic reactor in order for biological phosphorus uptake to occur. In the presence of dissolved oxygen, Poly-P bacteria must solubilize and degrade the PHA in order to release energy. Once released, the energy is captured and stored in phosphate bonds. In order to store as much phosphorus as possible, influent orthophosphate and released orthophosphate are absorbed from the bulk solution and used to produce high-energy phosphate bonds and polymers of phosphate that are stored as insoluble volutin granules.

Chapter 11

EBPR: Process Control

The advantages of enhanced biological phosphorus removal (EBPR) as compared to the use of chemical precipitation of phosphorus include:

- decreased operational costs
- decreased sludge production
- decreased sludge handling, dewatering, and disposal

These advantages can be achieved if the EBPR process is properly monitored and regulated to ensure that operational parameters are acceptable each day. Depending on the discharge requirement for total phosphorus (TP), EBPR may be used alone or in combination with other treatment processes including precipitation of phosphorus and tertiary treatment using filters (Table 11.1).

Table 11.1 Treatment Processes Available to Satisfy TP Discharge Requirements

Discharge Requirement (mg/L)	Treatment Processes
< 2.0	EBPR
< 1.0	EBPR Chemical precipitation Chemical precipitation and tertiary filtration

Discharge Requirement (mg/L)	Treatment Processes
< 0.5	EBPR Chemical precipitation Polymers Tertiary filtration
< 0.1	EBPR Chemical precipitation using several addition points Polymers Tertiary filtration

Daily monitoring of the EBPR process requires *in situ* and laboratory testing for acceptable biological, chemical, and physical parameters to ensure that the parameters are at or within the operational value or range of values that have been determined for each reactor. Each EBPR process is different and requires values that have been developed for proper operation and within ranges of values for changes in wastewater strength and composition and seasonal changes in temperature and inflow and infiltration.

Proper activity in the anaerobic/fermentative reactor includes:

- increase in orthophosphate concentration from influent to effluent
- decrease in ORP value from influent to effluent
- ORP values within the operating range from –100 mV to –250 mV
- production of PHA granules as observed with methylene blue staining

Proper activity in the aerobic react includes:

- decrease in soluble orthophosphate from influent to effluent
- increase in ORP value from influent to effluent
- ORP values within the operating range from +25 mV to +250 mV
- production of volutin granules as observed with basophilic staining
- phosphorus content in sludge > 5% on a dry weight basis

Improving EBPR performance

There are several parameters that should be monitored to ensure proper performance of each EBPR reactor. Some parameters apply to both reactors, while some parameters apply only to one reactor. The major parameters include: 1) population size of fermentative bacteria and Poly-P bacteria, 2) chemical oxygen demand (COD) loading, 3) cations, 4) glycogen-accumulating organisms 5) HRT, 6) loss of solids, 7) pH, and 8) temperature. Commonly recommended values or range of values for enhanced biological phosphorus removal are listed in Table 11.2.

Table 11.2 Commonly Recommended Values for EBPR

Parameter	Reactor	
	Anaerobic/fermentative	Aerobic
BOD-to-TP ratio	30:1 to 40:1	---
COD-to-TP ratio	> 45	---
Dissolved oxygen	---	0.5 to 1.0 mg/L in aerobic reactor effluent
HRT	0.5 to 2.0 hours	4 to 8 hours
MLSS	3,000 to 4,000	3,000 to 4,000
ORP	−100 to −200 mV	+100 to +300 mV
pH	8.0 to 8.5	7.3 to 8.0
Phosphorus content of sludge	---	> 5%
Temperature	20 to 25°C	20 to 25°C
VFA-to-TP ratio	---	> 7 influent

Population size of fermentative bacteria and Poly-P bacteria

An appropriate range of values for MLSS is needed in order to ensure an abundant population of fermentative bacteria and Poly-P bacteria. The bacteria enter the reactors through fecal waste and inflow and infiltration (I/I). As heterotrophic bacteria, their generation time is relatively short at 20 to 30 minutes. Without a process upset, overwasting of solids, or hydraulic washout from I/I, an appropriate range of values for MLSS must be maintained. If fermentative bacteria or Poly-P are needed, and natural "seeding" of the reactors through fecal waste and I/I does not provide the necessary MLSS for acceptable operation, the bacteria may be increased with the use of commercially available bio-augmentation products or the activated sludge from another activated sludge process that is practicing EBPR.

The relative abundance of bacteria in the activated sludge process is determined by the concentration of MLSS. If the process has TN and TP discharge requirements, then the concentration of MLSS must be large enough to support the growth of nitrifying bacteria and low enough to prevent overgrowth and die-off of Poly-P bacteria. At low concentrations of MLSS nitrifying bacteria will not be large enough to provide acceptable nitrification due to their long generation time (three to six days). At high concentrations of MLSS, Poly-P bacteria will be too large in population size due to their short generation time (20 to 30 minutes), resulting in the die-off and autolysis of many bacteria. Autolysis releases phosphorus and nitrogen to the bulk solution. Therefore, a range of concentrations for MLSS, for example 3000 to 4000 mg/L, should be established and maintained, and seasonally adjusted for swings in temperature. The range in concentrations for each process will be specific for that process.

COD loading

Because fatty acids are necessary for biological phosphorus release, an adequate organic loading must be provided to the anaerobic/fermentative reactor. A reduced or excess loading can lead to decreased activity of EBPR. A supplemental substrate that may be added to the anaerobic/fermentative reactor to increase COD loading is acetic acid. On-site fermenters may also be used to increase COD loading. Primary sludge and RAS may be used as on-site fermenters. Because cBOD testing cannot provide daily data on an as-needed basis for process control, chemical oxygen demand (COD) testing is used.

In addition to reduced or excess COD loading, significant swings in COD also adversely affect EBPR, the influent to the anaerobic/fermentative reactor should be equalized, and slug discharges should be prevented. An example of a sudden change in organic loading and HRT is the dilution of COD by I/I. For each EBPR process it is necessary to determine the optimal range of COD loading, and whether the range needs adjustment due to seasonal temperature changes in the wastewater.

Cations

There are three cations needed in the EBPR process to assist in phosphorus uptake and intracellular storage of phosphate. The cations are: calcium (Ca^{2+}), magnesium (Mg^{2+}), and potassium (K^+). Municipal wastewater typically contains an adequate supply of these cations for cBOD and total suspended solids (TSS) removal, but occasionally may be deficient in potassium for EBPR. Biological nutrient requirements for these micronutrients (given as a percentage of total cell mass) are:

- Calcium   1.4 percent
- Magnesium   0.5 percent
- Potassium   1.5 percent

Glycogen-accumulating bacteria

Glycogen is a multi-branched polysaccharide of glucose ($C_6H_{12}O_6$). It may contain up to 30,000 glucose units that serve as a form of energy storage.

Glycogen-accumulating organisms (GAO), specifically glycogen-accumulating bacteria (GAB) such as *Candidatus* and *Rhodocyclus*, use the same substrates as do Poly-P bacteria. However,

GAB do not participate in biological phosphorus removal beyond their cellular needs. In the anaerobic/fermentative reactor GAB also absorb the substrates more rapidly than Poly-P bacteria. GAB convert the absorbed substrate to glycogen instead of PHA.

By rapidly absorbing substrate more rapidly than Poly-P bacteria in the anaerobic/fermentative reactor, GAB decrease EBPR efficiency. However, GAB only dominate and outcompete Poly-P bacteria when the pH is < 7.2. Therefore, it is important to maintain operational conditions that favor the growth of Poly-P bacteria and disfavor the growth of GAB. The critical operational conditions that must be maintained in the anaerobic/fermentative reactor are pH > 7.2 and an MCRT > 20 days. At pH values > 7.2 GAB are sluggish. Since GAB have a slower growth rate than Poly-P bacteria, an MCRT > 20 days favors the growth of Poly-P bacteria.

HRT

Hydraulic retention time (HRT) is a critical operational parameter for proper performance of the anaerobic/fermentative reactor. A high retention time is needed for maximum fatty acid production and maximum phosphorus release. HRT values vary from plant to plant and may need to be extended to provide maximum phosphorus release.

Loss of solids

When solids are lost from the activated sludge process the following changes occur: 1) decrease in the number of bacteria in the EBPR process and mixed liquor resulting in decreased treatment efficiency, 2) possible permit violations for BOD, cBOD, and TSS, and 3) possible permit violations for TN and TP due to the presence of organic nitrogen and organic phosphorus in the solids. Loss of solids can occur through interruption of proper floc formation and poorly settling solids (Table 11.3) or production of fine solids (dispersed growth, colloids, and particulate material) (Table 11.4).

Table 11.3 Operational Conditions Associated with Poorly Settling Solids

Fats, oils, and grease (FOG)
Foam production and accumulation
Increase in percent MLVSS
Low dissolved oxygen concentration (0.8 mg/L for 10 or more consecutive hours)
Nutrient deficiency
Slug discharge of soluble cBOD
Undesired growth of filamentous organisms
Zoogloeal growth (viscous floc)
Young sludge age or MCRT

Table 11.4 Operational Conditions Associated with the Production of Fine Solids

Cell-bursting agents and surfactants
High/low pH
Salinity (total dissolved solids)
Septicity
Toxicity

Good settling solids are needed for EBPR. An increase in secondary clarifier capacity may be needed to obtain effective suspended solids removal. Good settling solids as determined through a 30-minute settleability of MLSS yield a sludge volume index of 90 to 120 mL/g.

pH

Biological phosphorus release (anaerobic/fermentative reactor) and biological phosphorus uptake (aerobic reactor) have different pH values that provide optimal performance. Acceptable range of pH for the anaerobic/fermentative reactor is from 8.0 to 8.5, while an acceptable range of pH for the aerobic reactor is from 7.3 to 8.0.

Temperature

Temperature influences enzymatic activity of bacteria, gas-transfer rates, and solids settleability. Below the optimum growth rate for a bacterium, the growth rate doubles approximately every 10°C until the optimum growth rate is reached. However, Poly-P bacteria are lower-range mesophiles and psychrophiles and predominate at ≤ 20°C (Table 11.5). GAB are mesophiles with an optimum growth rate at 25 to 32°C. Therefore, efficient EBPR occurs at temperatures ≤ 20°C due to the psychrophilic nature of Poly-P bacteria. However, rapid deterioration of EBPR can occur at any operational temperature due to loss of solids such as the undesired growth of the filamentous organisms *Microthrix parvicella*.

Table 11.5 Growth Ranges for Thermophilic, Mesophilic, and Psychrophilic Bacteria

Temperature Range (°C)	Bacterial Group		
	Thermophiles	Mesophiles	Psychrophiles
−15 to 12			X
10 to 50		X	
45 to 80	X		

Additional operational parameters affecting EBPR

Microthrix parvicella and Zoogloeal growth

There are two undesired biological conditions associated with activated sludge processes that practice EBPR: undesired growth of *Microthrix parvicella* and undesired growth of Zoogloeal organisms. *Microthrix parvicella* is a filamentous organism associated with production of viscous, chocolate-brown foam. It contributes to the development of poorly settling solids. It is associated with cold wastewater temperature ($< 16°C$) and FOG. Zoogloeal growth is the rapid and undesired proliferation of floc-forming bacteria such as *Zoogloea ramigera*. The bacteria produce a copious quantity of gelatinous material that accumulates in the floc particles, resulting in loss of floc density and entrapment of air and gas bubbles. Zoogloeal growth is associated with billowy white foam, and its growth is usually caused by a fermentative condition upstream of an aerobic reactor.

MLSS

To provide for an abundant and adequate population of Poly-P bacteria, 3,000 to 4,000 mg/L mixed liquor suspended solids (MLSS) should be maintained. If the concentration of MLSS is too high, many Poly-P bacteria will die due to the lack of adequate substrate. The lack of adequate substrate may be corrected by 1) decreasing the MLSS concentration or 2) adding a carbon source to the reactor.

Phosphorus loading

Inadequate phosphorus loading to the anaerobic/fermentative reactor reduces the efficiency of the EBPR process. An adequate phosphorus loading is necessary to promote the growth of Poly-P bacteria over glycogen-accumulating bacteria. The COD-to-TP ratio may be used to determine the presence of an adequate phosphorus loading. A ratio > 45 for the COD-to-TP is an indicator of an adequate phosphorus loading.

Secondary phosphorus release

Secondary phosphorus release occurs in the anaerobic/fermentative reactor when volatile fatty acids (VFA) are no longer available. The lack of VFA may be due to: 1) insufficient substrate entering the reactor, 2) a long HRT, 3) sluggish activity of the fermentative bacteria, and 4) the presence of nitrates.

Struvite

The development of struvite, magnesium ammonia phosphate ($MgNH_4PO_4 \cdot 6H_2O$), often is found in activated sludge processes that practice EBPR. Struvite is an insoluble white to yellow-white or brownish-white crystal. Deposits of struvite are responsible for several operational problems including:

- clogging of pipes and valves
- interference with analytical analyses
- reducing the operational life of equipment.

Abbreviations and Acronyms

A/O	Anoxic/Oxic
A²/O	Anaerobic/Anoxic/Oxic
AOB	Ammonium-oxidizing bacteria
BNR	Biological nutrient removal
BOD	Biochemical oxygen demand
°C	degrees Celsius
COD	Chemical oxygen demand
EBPR	Enhanced biological phosphorus removal
F/M	Food-to-microorganism ratio
FOG	Fats, oils, and grease
GAB	Glycogen-accumulating bacteria
GAO	Glycogen-accumulating organisms
HRT	Hydraulic retention time
IFFAS	Integrated fixed film-activated sludge
I/I	Inflow and infiltration
kcal	Kilocalorie
MCRT	Mean cell residence time
MLE	Modified Ludzack-Ettinger process
mV	Millivolt
MLVSS	Mixed liquor volatile suspended solids
cBOD	Carbonaceous biochemical oxygen demand
nBOD	Nitrogenous biochemical oxygen demand
NOB	Nitrite-oxidizing bacteria
ORP	Oxidation-reduction potential
PAO	Phosphorus-accumulating organisms
Poly-P	Phosphorus-accumulating bacteria

RAS	Return activated sludge
SBR	Sequencing batch reactor
SRT	Solids retention time
TA	Total ammonia/ammonium
TIN	Total inorganic nitrogen
TAN	Total ammonia/ammonium-nitrogen
TKN	Total Kjeldahl nitrogen
TP	Total phosphorus
TSS	Total suspended solids
UCT	University of Cape Town process
VFA	Volatile fatty acids
WETT	Whole effluent toxicity testing
μm	Micron

Glossary

Actinomycetes	a group of Gram-positive filamentous organisms including *Nocardia* that are responsible for the production of viscous chocolate-brown foam in activated sludge processes
acute	rapid and short-term impact
ammonification	production of ammonia from the degradation of organic nitrogen compounds
anaerobic	condition where cellular metabolism occurs in the absence of free molecular oxygen
anhydrous	without water
anoxic	condition where cellular metabolism occurs using nitrate or nitrite
assimilatory	to incorporate into cellular material
autolysis	splitting or opening up of bacteria cells upon death
autotrophic	organism that uses inorganic carbon (carbon dioxide) for cellular growth
bio-augmentation	the addition of culturally prepared cultures of archaea, bacteria, or fungi to improve treatment plant performance
biochemical	chemical reactions of living cells
cations	positively charged ions such as calcium (Ca^{2+}) and potassium (K^+)
chronic	slow and long-term impact
coagulants	metal salt such as ferric chloride ($FeCl_3$) used to capture and thicken solids

colloids	large and complex molecules such as proteins that do not dissolve in wastewater
Cyanobacteria	a large group of blue-green algae, sometimes referred to as blue-green bacteria
deamination	removal of amine ($-NH_2$) group from organic nitrogen compounds such as amino acids
detritus	loose materials such as rock fragments or organic particles
dissimilatory	not incorporated
electron acceptor	an inorganic molecule such as free molecular oxygen or nitrate or an organic molecule that accepts freed electrons from broken chemical bonds of degraded substrate
endogenous	the solubilization and degradation of stored food inside and outside cells during periods of low substrate loading
eutrophication	rapid enrichment of bodies of waters with aquatic flora and fauna
facultative anaerobe	an anaerobic organism that is capable of using free molecular oxygen
fermentation	degradation of organic substrate in the absence of an inorganic electron resulting in the production of a mixture of acids and alcohols
habitat	where an organism lives
heterotrophic	an organism that is dependent upon another organism or wastes of another organism (organic compounds) for carbon and energy for cellular growth
hydrolysis	splitting of the chemical bonds in complex insoluble molecules such as starch, resulting in the production of simplistic soluble molecules such as glucose
indigenous	naturally occurring, not introduced

inorganic   a compound or ion that does not contain carbon and hydrogen

leachate   a solution resulting from the drainage of solids

metabolism   any form of cellular activity

methemoglobinemia also known as blue-baby syndrome. Disease caused in infants by the consumption of nitrate-contaminated water.

obligate   required or strict

organic   molecule containing carbon and hydrogen

oxidation state   the electrical charge (positive, negative, or neutral) of an atom

pathogen   disease-causing virus or organism

phytic acid   large and complex cyclic acid containing numerous phosphorus groups

polymer   a large molecule composed of many repeating units of the same molecule

psychrophilic   an organism that grows in cold environments

recalcitrant   difficult to degrade or slowly degrading compound

respiration   degradation of substrate

septicity   a condition having an oxidation-reduction potential < -100 mV where incomplete degradation of substrate occurs and malodors are produced

salinity   dissolved salt content of wastewater

thermophilic   an organism that grows in a hot environment

volutin   intracellular granule of stored polyphosphate

Bibliography

Allen, J. E. 1984. Elevated nitrite occurrence in biological wastewater treatment systems. *Water Science Tech.* 17: 409–419.

Arora, M. L., et al. 1985. Technology evaluation of sequencing batch reactors. *J. WPCF.* 57(8): 867–875.

Balmelle, B. K., et. al. 1992. Study of reactors controlling nitrite build-up in biological processes for water nitrification. *Water Science Tech.* 26: 1017–1025.

Bond, P. L., et al. 1999. Identification of some of the major groups of bacteria in efficient and non-efficient biological phosphorus removal activated sludge systems. *Appl. Environ. Microbiol.* 65(9): 4077–4084.

Chen, J. S, et al. 1993. Competition between polyphosphate and polysaccharide accumulating bacteria in enhanced biological phosphate removal systems. *Water Res.* 27(7): 203–211.

Chen, W. L., and J. N. Jensen. 2001. Effect of chlorine demand on the breakpoint curve: model development, validation with nitrite, and application to municipal wastewater. *Water Environ. Res.* 73(6): 721.

Choi, E., et. al. 1998. Temperature effects on biological nutrient removal system with weak municipal wastewater. *Water Sci. Technol.* 37(9): 219–226.

Comeau, Y., et al. 1986. Biochemical model for enhanced biological phosphorus removal. *Water Res.* 20(12): 511–512.

Constantin, H., and M. Fick. 1977. Influence of c-sources on denitrification rate of a high nitrate concentrated industrial wastewater. *Water Res.* 31(7): 583–589.

Correll, D. L. 1998. The role of phosphorus in the eutrophication of receiving waters: a review. *J. Environ. Qual.* 27: 261–266.

Filipe, C. D. M., et al. 2001. Effects of pH on the rates of aerobic metabolism of phosphate-accumulating and glycogen-accumulating organisms. *Water Environ. Res.* 73(2): 213–222.

Filipe, C. D. M., et al. pH as a key factor in the competition between glycogen-accumulating and organisms and phosphorus-accumulating organisms. *Water Environ. Res.* 73(2): 223–232.

Focht, D. D., and A. C. Chang. 1975. Nitrification and denitrification processes related to waste treatment. *Adv. Appl. Microbiology.* 19: 153–186.

Hellinga, C., et al. 1998. The SHARON process: an innovative method for nitrogen removal from ammonium rich wastewater. *Water Sci. Technol.* 37: 135–142.

Helmer, C. and S. Kunst. 1998. Low temperature effects on phosphorus release and uptake by microorganisms in EBPR plants. *Water Sci. Technol.* 37(4–5): 531–539.

Kristensen, G. H., et al. 1994. Settling characteristics of activated sludge in Danish treatment plants with biological nutrient removal. *Water. Sci. Technol.* 29(7): 157–165.

Kornaros, M., et al. 1996. Kinetics of denitrification by *Pseudomonas denitrificans* under growth conditions limited by carbon and/or nitrate or nitrite. *Water Environ. Res.* 68(5): 934–935.

Li, J., et al. 2005. Technique for biological phosphorus removal. *Pollution Engineering.* 37: 14–17.

Mamais, D., and D. Jenkins. 1992. The effects of mean cell residence time and temperature on enhanced biological phosphorus removal. *Water Sci. Technol.* 26(5–6): 955–965.

Metcalf & Eddy. 1991. *Wastewater Engineering; Treatment, Disposal, and Reuse*, 3rd Edition. New York: McGraw-Hill.

Mino, T., et al. 1998. Microbiology and biochemistry of the enhanced biological phosphate removal process. *Water Res.* 32(11): 3193–3207.

Mulkerrins, D., et al. 2004. Parameters affecting biological phosphate removal from wastewater. *Environ. International.* 30: 249–259.

Nurse, G. R. 2003. Denitrification with methanol: microbiology and biochemistry. *Water Res.* 14(5): 531–537.

Randall, C. W., and R. W. Chapin. 1997. Acetic acid inhibition of biological phosphorus removal. *Water Environment Research.* 69(5): 955–960.

Satoh, H., et al. 1992. Uptake of organic substrates and accumulation of polyhydroxy-alkanoates linked with glycolysis of intracellular carbohydrates under anaerobic conditions in the biological excess phosphate removal processes. *Water Sci. Technol.* 26(5–6): 933-942.

Sawyer, C., et al. 1994. *Chemistry for Environmental Engineering,* 4th Edition. New York: McGraw-Hill.

Tang, N. H., et al. 1992. OSAR parameters for toxicity of organic chemicals to *Nitrobacter*. *Jour. of Environ. Eng.* 118(1).

Thongchai, P., et al. 2003. Temperature effect on microbial community of enhanced biological phosphorus removal system. *Water Res.* 37: 409–415.

US EPA. 1993. *Manual: Nitrogen Control,* EPA Office of Research and Development, Cincinnati, Ohio. EPA/625/R-93/010, Office of Water. Washington, DC.

Van Loosdrecht, et al. 1998. Microbiological conversions in nitrogen removal. *Water Science Tech.* 38: 1–7.

Water Environment Federation. 2011. *Nutrient Removal: WEF Manual of Practice No. 34.* New York: McGraw-Hill.

Water Environment Federation. 2001. *Natural Systems for Wastewater Treatment,* 2nd Edition. Alexandria, Virginia.

Water Environment Federation. 1998. *Biological and Chemical Systems for Nutrient Removal, Special Publication.* Alexandria, Virginia.

Wu, Q., et al. 2005. Biological phosphate uptake and release: effect of pH and magnesium ions. *Water Environ. Res.* 77.

www.ingramcontent.com/pod-product-compliance
Lightning Source LLC
Chambersburg PA
CBHW070725220326
41598CB00024BA/3300